The Unofficial

Grades 1-4
Ages 6+

POKEMON

Coloring Math Book

Multiplication and Division

Part 1

A Complete Guide to Master

1600+ Tasks and Word Problems

with

Word Search, Mazes, Comics,

CogAT® test prep, and more!

Copyright © 2019 STEM mindset, LLC. All rights reserved.

STEM mindset, LLC 1603 Capitol Ave. Suite 310 A293 Cheyenne, WY 82001 USA
www.stemmindset.com   info@stemmindset.com

This book is not authorized, endorsed or affiliated with The Pokemon Company, Satoshi Tajiri, Game Freak or Nintendo. All Pokemon references are used with the Fair Use Doctrine and are not meant to imply that this book is a Pokemon product for advertising or other commercial purpose. All aspects of the game including characters, their names, locations, and ther features of the game within this book are trademarked and owned by their espective owners. This book is a work of fiction. Names, characters, places, and incidents are either the product of authors imagination or are used fictitiously, and any resemblance to actual persons, living or dead, business establishments, events, or locales is entirely coincidental.

The purchase of this material entitles the buyer to reproduce worksheets and activities for classroom use only – not for commercial resale. Reproduction of these materials for an entire school or district is strictly prohibited. No part of this book may be reproduced (except as noted before), stored in retrieval system, or transmitted in any form or by any means (mechanically, electronically, photocopying, recording, etc.) without the prior written consent of STEM mindset, LLC.

First published in the USA 2019.   ISBN 9781948737593

Table of Contents

Practice: Understanding Division by 2s. Word Problem. Coloring	7
Practice: Understanding Division by 2s, 4s, 5s	8
Practice: Understanding Division by 2s, 3s, 4s, 6s	9
Practice: Understanding Division by 2s, 3s, 5s, 7s	10
Practice: Understanding Division by 2s, 4s, 8s. Word Problem. Coloring	11
Maze. Word Search. Coloring	12
Practice: Understanding Division by 2s, 3s, 6s. Word Problem. Coloring	13
Practice: Understanding Division by 2s, 4s, 5s, 10s	14
Dividing a Number by 1 or by Itself. Multiplying by 0. Coloring	15
Understanding Multiplication as Repeated Addition. Coloring	16
Understanding Multiplication Using Arrays. Word Problem. Coloring	17
Understanding Multiplication Using Arrays. Factors. Product. Word Problem. Coloring	18
Practice: Multiplying by 2s and by 3s. Word Problem. Coloring	19
Maze. Word Search. Coloring	20
Understanding Division and Multiplication by 2s. Word Problem. Coloring	21
Understanding Division by 5s. Word Problem. Coloring	22
Practice: Dividing by 2s, 3s, 4s	23
Practice: Dividing by 2s, 5s, 6s, 8s	24
Practice: Dividing by 2s, 3s, 4s, 7s. Word Problem. Coloring	25
Practice: Dividing by 2s, 4s, 5s, 9s, 10s	26
Maze. Word Search. Coloring	27
Practice: Multiplying and Dividing by 2s Using Arrays. Word Problem. Coloring	28
Practice: Multiplying and Dividing by 2s Using Arrays. Word Problem. Coloring	29
Practice: Multiplying by 2s in Columns. Maze. Coloring	30
Practice: Multiplying and Dividing within 20. Word Problem. Coloring	31

Table of Contents

Practice: Multiplying and Dividing within 20. Word Problem. Coloring	32
Practice: Multiplying and Dividing within 20. Word Problem. Coloring	33
Practice: Finding Missing Factors. Maze. Coloring	34
Practice: Multiplying within 20. Word Problem	35
Maze. Word Search. Coloring	36
Practice: Multiplying and Dividing by 3s Using Arrays. Word Problem. Coloring	37
Practice: Multiplying and Dividing by 3s Using Arrays. Word Problem. Coloring	38
Practice: Multiplying by 3s in Columns. Maze. Coloring	39
Practice: Multiplying and Dividing within 30. Word Problem. Coloring	40
Practice: Multiplying and Dividing within 30. Coloring	41
Practice: Multiplying and Dividing within 30. Word Problem. Coloring	42
Practice: Finding Missing Factors. Maze. Coloring	43
Practice: Multiplying within 30. Word Problem	44
Maze. Word Search. Coloring	45
Practice: Multiplying. Finding Missing Factors. Dividing with Remainders within 5	46
Practice: Dividing with Remainders within 7. Word Problem. Coloring	47
Practice: Multiplying and Dividing within 20. Finding Missing Divisors	48
Practice: Dividing with Remainders within 10. Coloring	49
Practice: Puzzle. Finding Missing Factors or Dividends. Coloring	50
Maze. Word Search. Coloring	51
Practice: Puzzle. Finding Missing Factors or Dividends. Coloring	52
Practice: Comparing. Understanding Order of Operations. Word Problem. Coloring	53
Practice: Puzzle. Order of Operations. Coloring	54
Practice: Finding Missing Factors or Products	55

Table of Contents

Practice: Puzzle. Finding Missing Divisors. Order of Operations. Word Problem. Coloring	56
Practice: Multiplying. Dividing. Finding Missing Factors. Dividing with Remainders within 13. Coloring	57
Maze. Word Search. Coloring	58
Practice: Multiplying. Dividing. Finding Missing Factors. Dividing with Remainders within 17. Word Problem. Coloring	59
Practice: Dividing with Remainders within 15. Word Problem. Coloring	60
Practice: Finding Missing Factors or Products	61
Practice: Dividing with Remainders within 18. Word Problem. Coloring	62
Practice: Finding Missing Factors or Products	63
Practice: Dividing with Remainders within 19. Word Problem. Coloring	64
Maze. Word Search. Coloring	65
Practice: Dividing within 30. Multiplying by 2s-9s. Word Problem. Coloring	66
Practice: Dividing with Remainders within 20. Word Problem. Coloring	67
Practice: Multiplying and Dividing. Word Problem. Coloring	68
Practice: Multiplying and Dividing by 4s Using Arrays. CogAT® Test Prep: Equation Building. Coloring	69
Practice: Multiplying and Dividing by 4s Using Arrays. CogAT® Test Prep: Equation Building. Coloring	70
Practice: Multiplying by 4s in Columns. Maze. Coloring	71
Practice: Multiplying and Dividing within 40. CogAT® Test Prep: Quantitative Relations	72
Practice: Multiplying and Dividing within 40. CogAT® Test Prep: Quantitative Relations	73
Practice: Multiplying and Dividing within 40. CogAT® Test Prep: Quantitative Relations. Coloring	74
Practice: Finding Missing Factors. Maze. Coloring	75

Table of Contents

Practice: Multiplying within 40. Word Problem	76
Maze. Word Search. Coloring	77
Practice: Multiplying and Dividing by 5s Using Arrays. CogAT® Test Prep: Quantitative Relations. Coloring	78
Practice: Multiplying and Dividing by 5s Using Arrays. CogAT® Test Prep: Quantitative Relations. Coloring	79
Practice: Multiplying by 5s in Columns. Maze. Coloring	80
Practice: Multiplying and Dividing within 50. CogAT® Test Prep: Number Series. Coloring	81
Practice: Multiplying and Dividing within 50. CogAT® Test Prep: Number Series. Coloring	82
Practice: Multiplying and Dividing within 50. CogAT® Test Prep: Equation Building. Coloring	83
Practice: Finding Missing Factors. Maze. Coloring	84
Practice: Multiplying within 50. Word Problem	85
Maze. Word Search. Coloring	86
Practice: Multiplication Strategies. CogAT® Test Prep: Equation Building	87
Practice: Multiplication Strategies. Coloring	88
Practice: Understanding Division. CogAT® Test Prep: Equation Building. Coloring	89
Practice: Order of Operations	90
Practice: Understanding Division	91
Maze. Word Search. Coloring	92
Practice: Finding Missing Factors or Products	93
Practice: Understanding Multiplication. Order of Operations	94
Practice: Multiplying by 2s-9s	95
Practice: Dividing by 2s-9s	96

Table of Contents

Maze. Word Search. Coloring   97

Answers   98

EVERYBODY'S A CRITIC -
And if you're one, too, we want You!

Speak up, write on, and let your voice be heard! We want to know what you, parents and users, really think – share your feedback on info@stemmindset.com or www.amazon.com!

Kids learn and stay engaged, thanks to puzzles, coloring, mazes, word search tasks, along with challenging math problems. These methods help kids understand math concepts, master math skills, and even help struggling students gain the confidence to improve their math comprehension and testing.

These activities are perfect for daily practice, morning work, homework, math centers, early finishers, test preparation, assessment, or for high school students struggling with fractions.

They will work great in 1st Grade to challenge students. They are perfect in 2nd Grade to understand and master multiplication and division within 50, but they also might work in 3rd-5th Grades as a review or for those who are struggling with math.

Hi. I'm Sunny. For me, everything is an adventure. I am ready to try anything, take chances, see what happens - and help you try, too! I like to think I'm confident, caring and have an open mind. I will cheer for your success and encourage everyone! I'm ready to be a really good friend!

I've got a problem. Well, I've always got a problem. And I don't like it. It makes me cranky, and grumpy, impatient and the truth is, I got a bad attitude. There. I said it. I admit it. And the reason I feel this way? Math! I don't get it and it bums me out. Grrrr!

Not trying to brag, but I am the smartest Brainer that ever lived - and I'm a brilliant shade of blue. That's why they call me Smarty. I love to solve problems and I'm always happy to explain how things work - to help any Brainer out there! To me, work is fun, and math is a blast!

I scare easily. Like, even just a little ...Boo! Oh wow, I've scared myself! Anyway, they call me Pickles because I turn a little green when I get panicky. Especially with new stuff. Eek! And big complicated problems. Really any problem. Eek! There, I did it again.

Hi! Name's Pepper. I have what you call a positive outlook. I just think being alive is exciting! And you know something? By being friendly, kind and maybe even wise, you can have a pretty awesome day every day on this amazing planet.

A famous movie star once said, "I want to be alone." Well, I do too! I'm best when I'm dreaming, thinking, and in my own world. And so, I resist! Yes, I resist anything new, and only do things my way or quit. The rest of the Brainers have math, but I'd rather have a headache and complain. Or pout.

1. <u>Divide</u>.

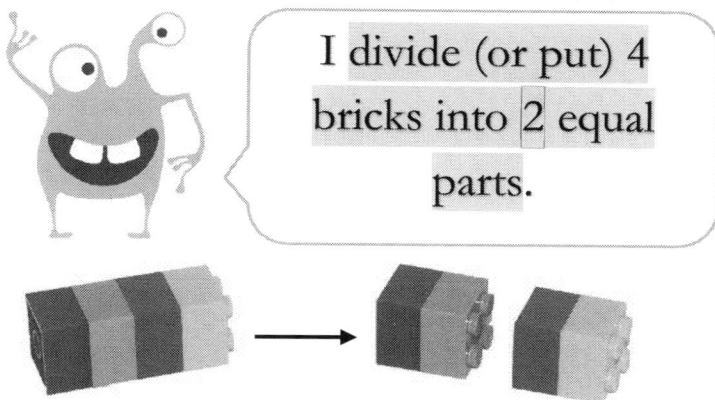

2. I saw 4 flowers on each of 2 flowes beds on the street of Floaroma town. <u>How many flowers</u> did I see on one flower bed?

$4 \div 2 = \_\_$

Answer: _____

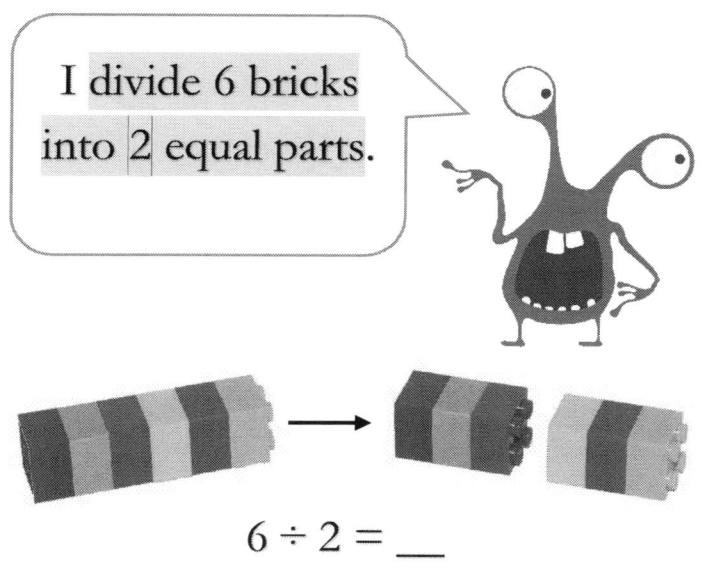

$6 \div 2 = \_\_$

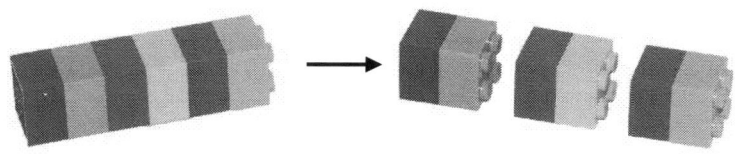

$6 \div 3 = \_\_$

1. <u>Divide</u>.

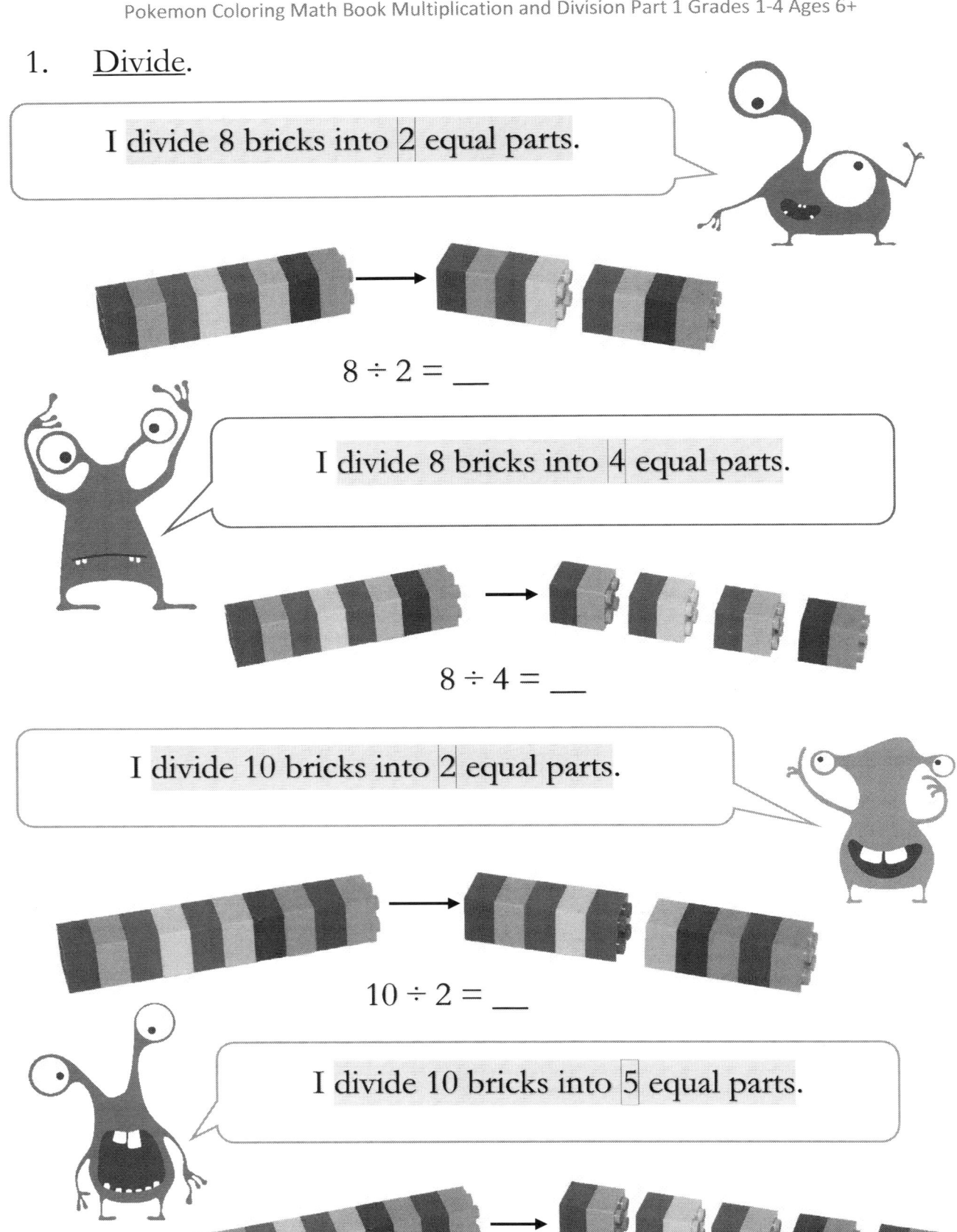

I divide 8 bricks into 2 equal parts.

$8 \div 2 = \underline{}$

I divide 8 bricks into 4 equal parts.

$8 \div 4 = \underline{}$

I divide 10 bricks into 2 equal parts.

$10 \div 2 = \underline{}$

I divide 10 bricks into 5 equal parts.

$10 \div 5 = \underline{}$

1. <u>Divide</u>.

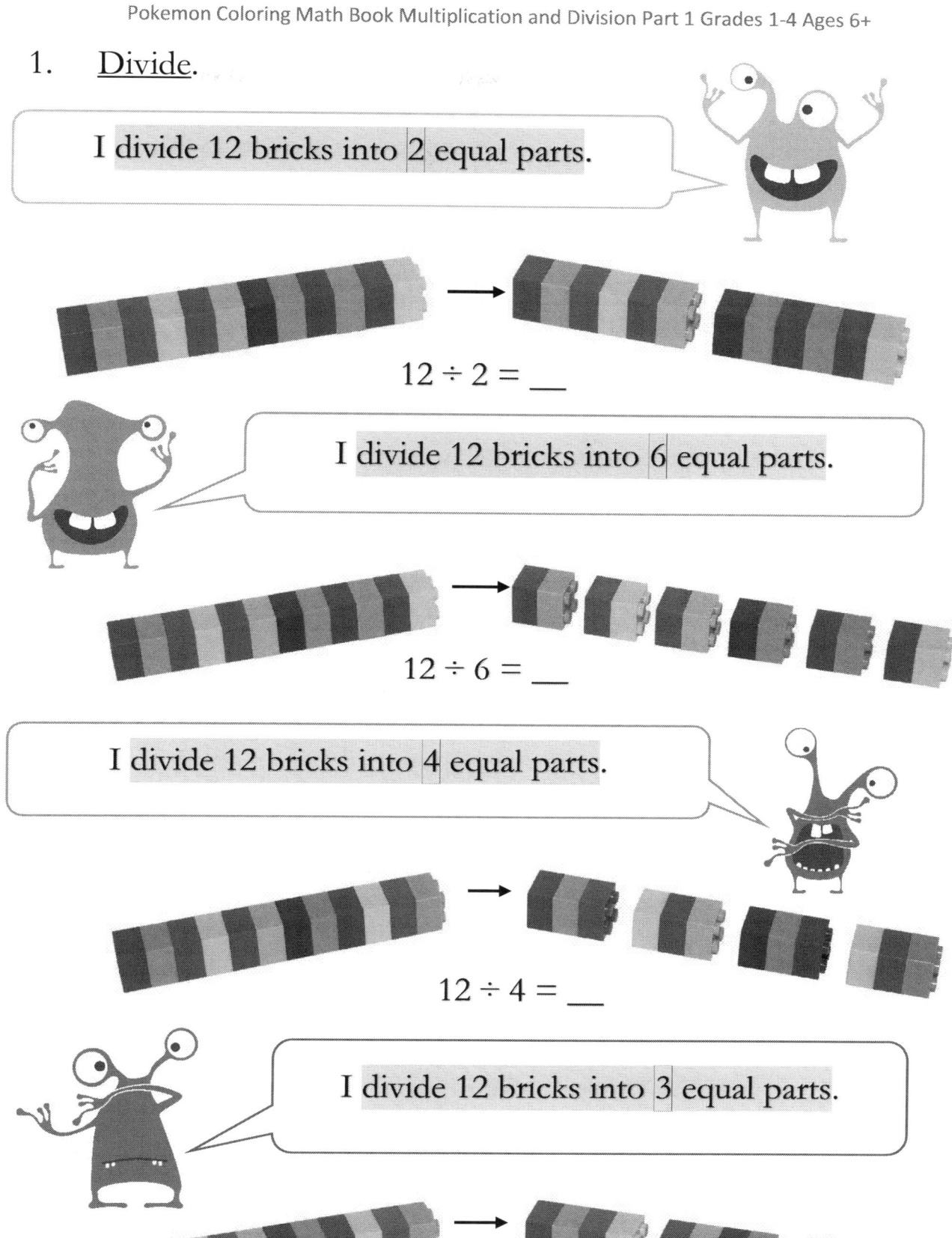

I divide 12 bricks into 2 equal parts.

$12 \div 2 = \_\_$

I divide 12 bricks into 6 equal parts.

$12 \div 6 = \_\_$

I divide 12 bricks into 4 equal parts.

$12 \div 4 = \_\_$

I divide 12 bricks into 3 equal parts.

$12 \div 3 = \_\_$

1. <u>Divide</u>.

I divide 14 bricks into 2 equal parts.

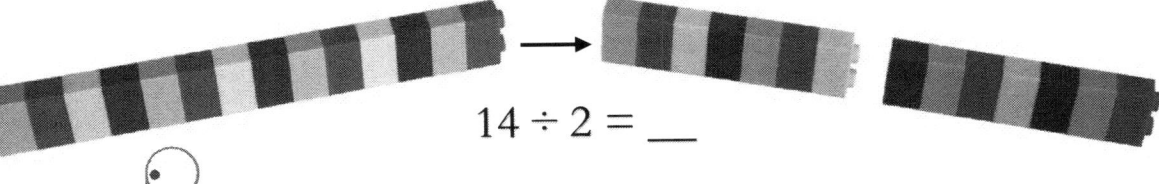

$14 \div 2 = \_\_$

I divide 14 bricks into 7 equal parts.

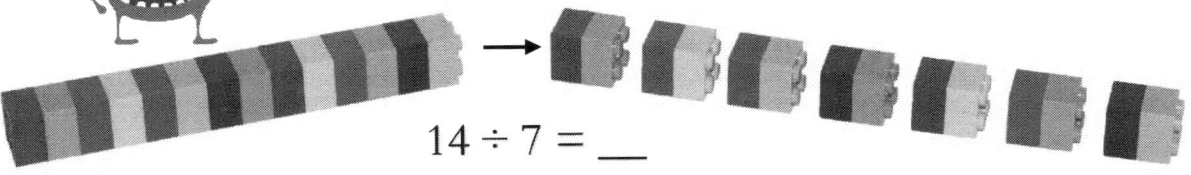

$14 \div 7 = \_\_$

I divide 15 bricks into 3 equal parts.

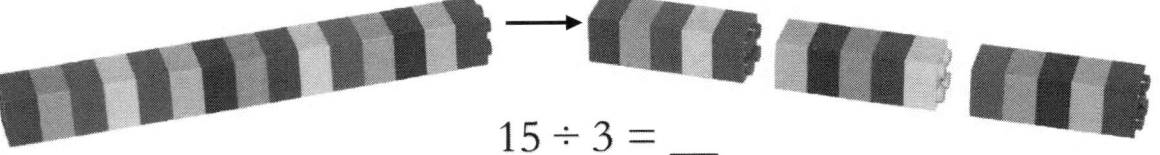

$15 \div 3 = \_\_$

I divide 15 bricks into 5 equal parts.

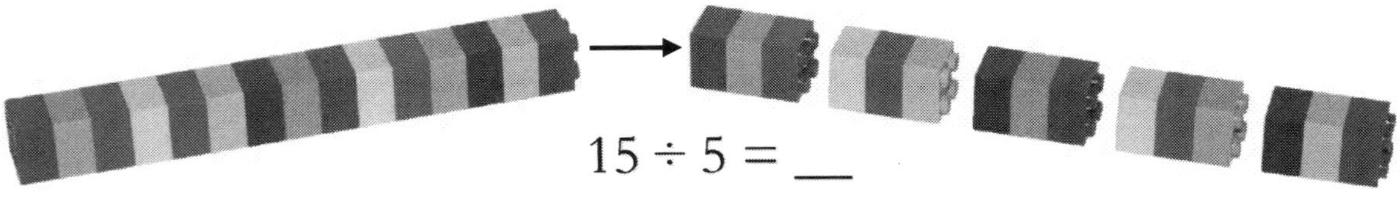

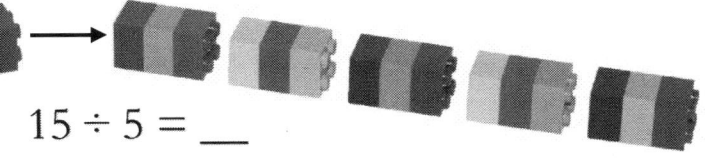

$15 \div 5 = \_\_$

1. <u>Divide</u>.

I divide 16 bricks into 2 equal parts.

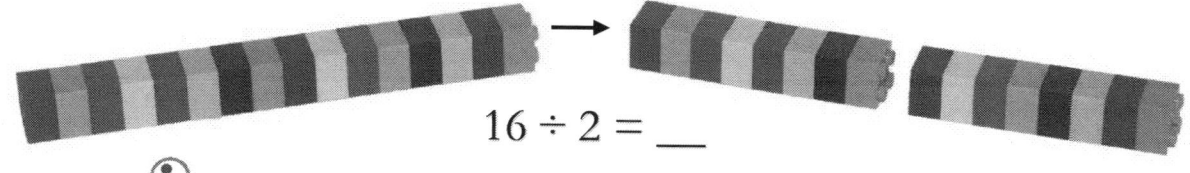

16 ÷ 2 = ___

 I divide 16 bricks into 4 equal parts.

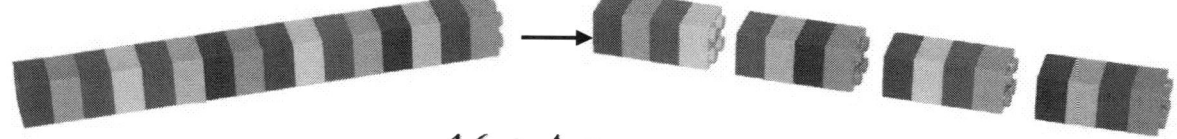

16 ÷ 4 = ___

I divide 16 bricks into 8 equal parts. Hate math!

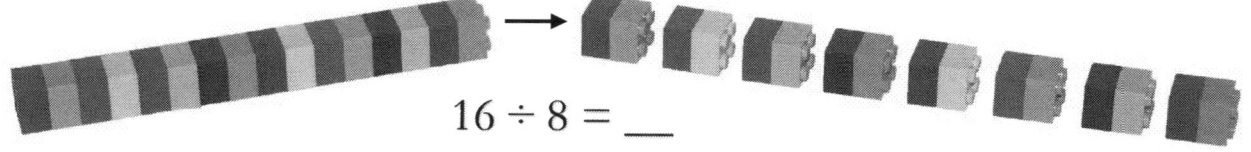

16 ÷ 8 = ___

2. Combee had 15 pounds of honey in the jars. Each jar contained 3 pounds. <u>How many jars</u> were there?

Answer: _____

1. <u>Get</u> Pikachu.

2. <u>Find</u> and <u>circle</u> or <u>cross out</u> the words.

However, as the time passes, division seems to fit me more and more… But, fellers, I still prefer mazes 😊.

D X P D T V Q A K Q Y V S
A Y L N S Q M C K T O T C
N G W C E M A P I L R W L
G Z R U H T M L Y E R H Q
E S Z E T A A X N H K C F
R M N A D N R G W V I H X
O V H X O E T M B S R V C
U K N S S H E K E R L G I
S O R L I Z A R D L I K E
Y E I M J C N I B A E Q A
P W F M D S K G I C Y O L
F M M W U G Z E T I B S N
T A S H W J W K U G E Y A
L U F E C A E P Z A W J I
D D S H U G P C A M Q J G

ASH
BREEDER
STRENGTH
PERSONALITY
LIZARDLIKE
CHARMELEON
DANGEROUS
PEACEFUL
MAGICAL
ATTACK

1. <u>Divide</u>.

I divide 18 bricks into 2 equal parts.

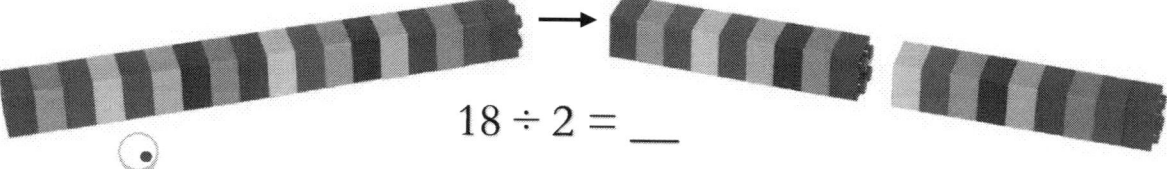

18 ÷ 2 = ___

I divide 18 bricks into 3 equal parts.

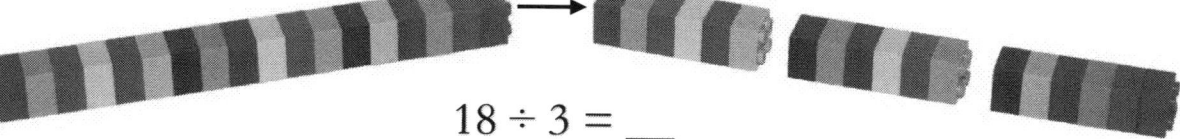

18 ÷ 3 = ___

I divide 18 bricks into 6 equal parts.

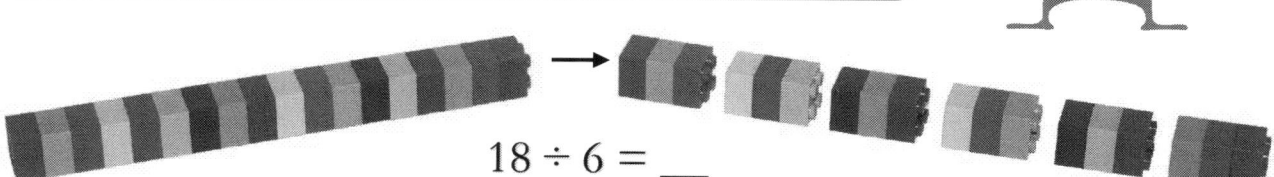

18 ÷ 6 = ___

2. Charmeleon's burning tail swang 16 times during the 4 of his attacks respectively. <u>How many times</u> did his tail swing during one attack?

Answer: _____

1. <u>Divide</u>.

I divide 20 bricks into 4 equal parts.

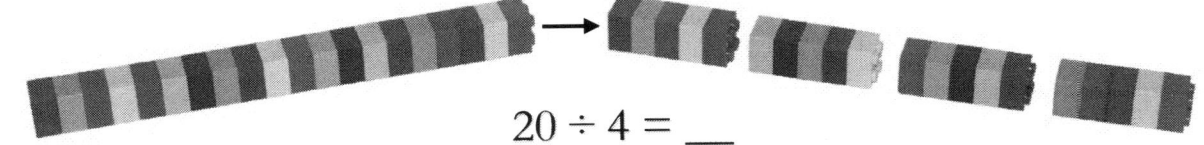

$20 \div 4 = \_\_$

I divide 20 bricks into 5 equal parts.

$20 \div 5 = \_\_$

I divide 20 bricks into 2 equal parts.

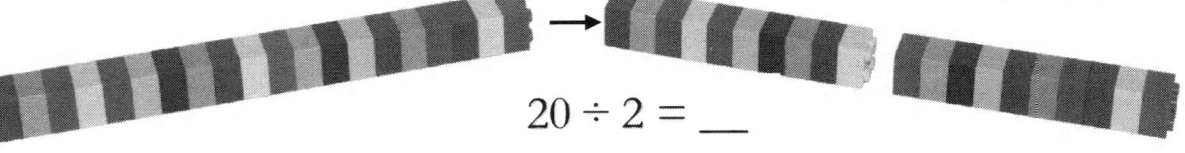

$20 \div 2 = \_\_$

I divide 20 bricks into 10 equal parts.

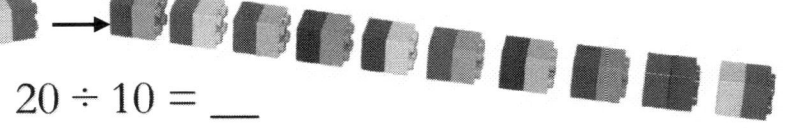

$20 \div 10 = \_\_$

Pokemon Coloring Math Book Multiplication and Division Part 1 Grades 1-4 Ages 6+

"I divide 15 bricks into 5 equal parts. Hate math!"

 "I will circle 15 bricks: each circle has only one (1) brick since I divide 15 by 15."

"What's all the fuss about, Pickles? A fella's got to learn multiplication sometime…"

$15 \div 15 = 1$

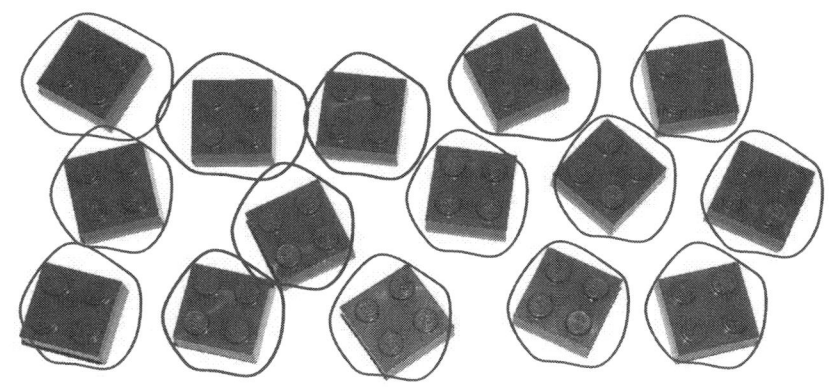

"Aha… any number divided by itself equals 1: $2 \div 2 = 1$; $85 \div 85 = 1$; $1 \div 1 = 1$."

"Any number multiplied by 1 equals itself since we take only 1 group of this number."

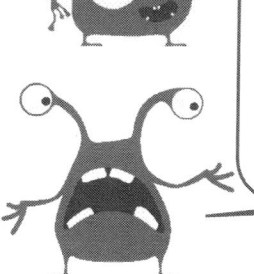

"Brilliant! Then, any number multiplied by 0 equals 0. Here, I take 0 groups of 11 bricks ☺."

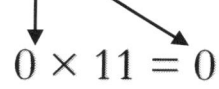
$0 \times 11 = 0$

$1 \times 11 = 11$

www.stemmindset.com © 2019 STEM mindset, LLC 15

1. <u>Add or multiply.</u>

Take the bricks or cubes and arrange them so that each row has 6 bricks and each column has 2 bricks.

<u>How many bricks</u> do you have in all? _____.

I know, I know! <u>Find</u> the sum of all the bricks in rows:

$6 + 6 =$ ___.

Or I have a 2 by 6 array of bricks:

$2 \times 6 =$ ___.

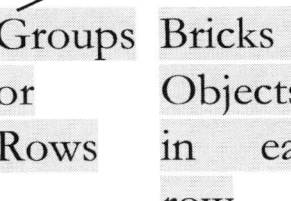

Yeah, Sunny! Math is the only language that makes any sense to me!

| Groups or Rows | Bricks or Objects in each row | Total in all rows or all bricks |

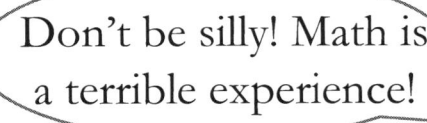

Don't be silly! Math is a terrible experience!

I would multiply.

The product of 2 bricks per column and 5 bricks per row equals: $2 \times 5 = 10$.

Then, we add 2 more bricks: $10 + 2 =$ ___.

Algorithms! It's boring! Ok, I am SURE it's easier to find the sum of all the bricks in columns:

2 + 2 + 2 + 2 + 2 + 2 = __.

Aha… A 6 by 2 array:

$6 \times 2 =$ __.

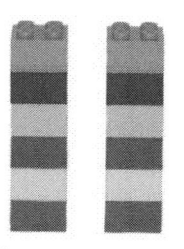

Oh, boy! What is multiplication? Array? Product? Pikachu?!

1. Zubat has $\boxed{2}$ wings. I met $\boxed{3}$ Zubats. <u>How many wings</u> did I see?

Answer: _____.

Multiplication is a repeated addition: take, for example $a \times b =$ you add "b" "a" times:

$$a \times b = \underbrace{b + b + b + \ldots + b}_{a \text{ times}}$$

Don't worry! I will make it easier! Look, I have some bricks: this is the <u>row</u>, this is the <u>column</u>. An array is formed by putting or arranging bricks into rows and columns.

Row 1 →
Row 2 →

Aha… this is a 2 by 5 array. It has 2 rows and 5 columns. So, I count rows first, right?

And I write it as $2 \times 5 =$ ___

If I make an addition number sentence, I need to add 5 bricks 2 times: $5 + 5 =$ ___

2 rows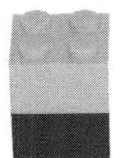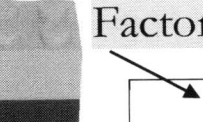

Factor Factor Product

$2 \times 5 =$ ___

$5 + 5 =$ ___

5 columns

1. Each of 3 Jigglypuffs sang 2 songs to lull their enemies to sleep. <u>How many songs</u> did they sing in all?

Answer: _____.

5 rows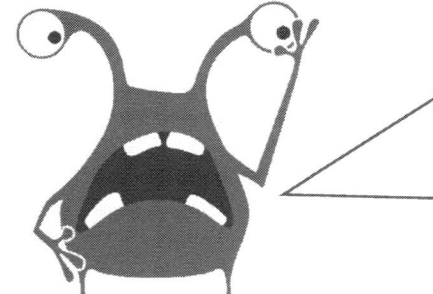

2 columns

Ok, I made a 5 by 2 array of bricks:

I have 5 rows and 2 columns. Since rows come first, I write:

$5 \times 2 =$ ___

Or I add 2 bricks 5 times:

$2 + 2 + 2 + 2 + 2 =$ ___

5 and 2 are factors.

18 © 2019 STEM mindset, LLC   www.stemmindset.com

1. <u>Multiply, add,</u> and <u>draw</u> arrows to <u>connect</u> the addition and multiplication number sentences of the same value and the picture of the bricks.

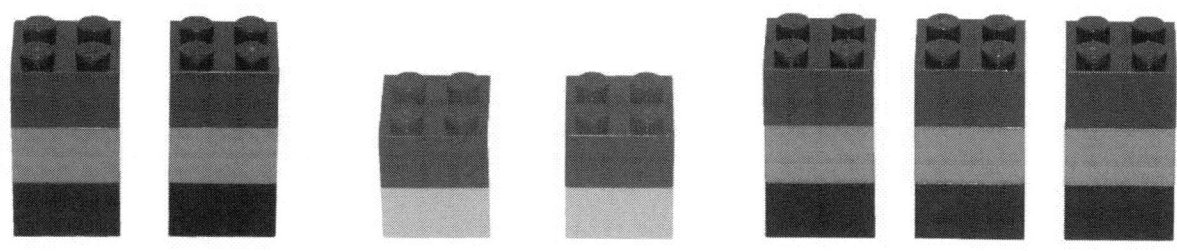

3 × 2 = __   2 × 2 = __   3 × 3 = __   2 + 2 = __

3 + 3 + 3 = __   2 + 2 + 2 = __

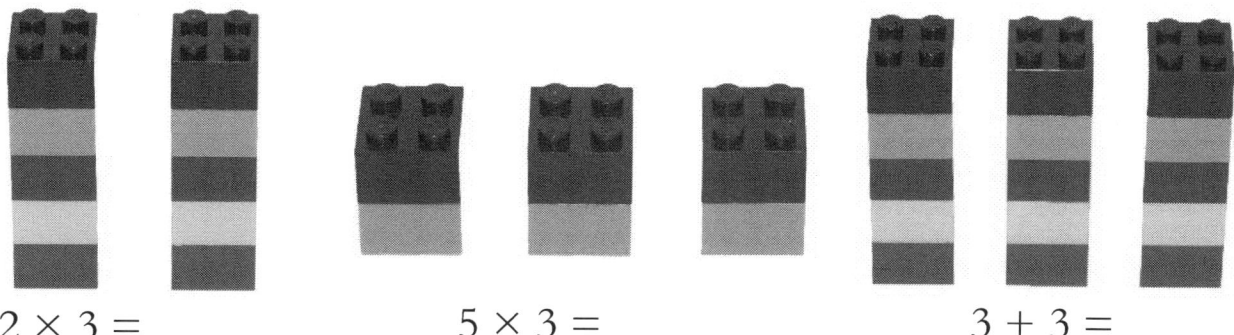

2 × 3 = __   5 × 3 = __   3 + 3 = __

5 × 2 = __   3 + 3 + 3 + 3 + 3 = __   2 + 2 + 2 + 2 + 2 = __

2. One Pokemon hatches in one egg. I have 5 eggs. <u>How many Pokemon</u> will I hatch?

Answer: _____

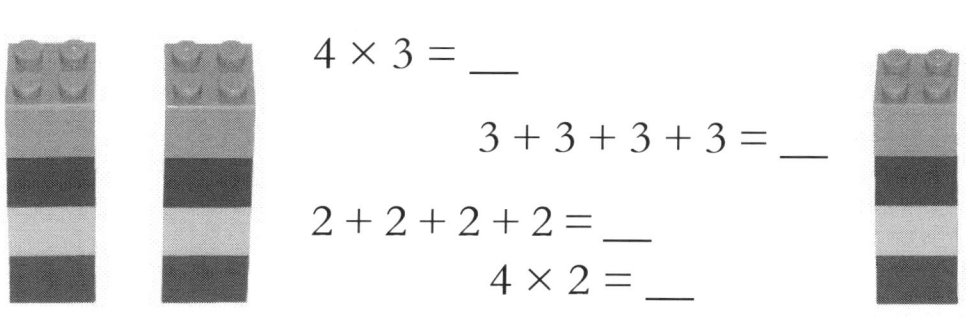

4 × 3 = __

3 + 3 + 3 + 3 = __

2 + 2 + 2 + 2 = __

4 × 2 = __

1. <u>Get</u> Pikachu.

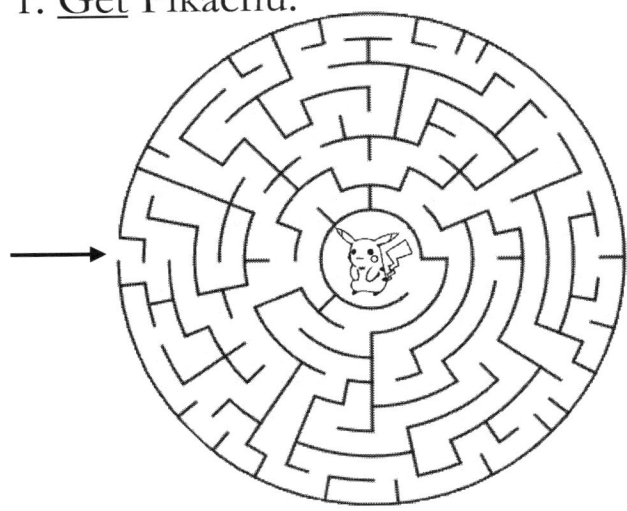

2. <u>Find</u> and <u>circle</u> or <u>cross out</u> the words.

```
Q M T P T R O T S L N I
W G U X A O A V P O P X
O O M S E R Y M I O X K
V F L U H A A T N C U J
M Q D L O R O S H A O L
J K J G A P O S P T P Y
M B C B K H I O S N P D
R G N M H F S T M E O N
V W P V Y Z I J O T N E
Z S W L H N J I E X E G
T J L O G P O I S O N E
J E Y E R M U M O M T L
J X R W P V T Z J O P O
E S N E F E D N P H I B
C L E F A I R Y Q J Y B
J F M J B P B R I H R Z
D A L B Z R Z F P V W P
```

CLEFAIRY

LEGEND

PARAS

MUSHROOM

DEFENSE

POTION

OPPONENT

TENTACOOL

SHALLOW

STINGERS

JELLYFISH

POISON

 How many bricks do I have?

I see [10] bricks: 5 × 2 = 10

I made [2] rows.

How many bricks are there per row?

 I see 2 rows with 5 bricks per row:

10 ÷ 2 = 5

Dividend = number of bricks in all

Quotient = number of bricks per row

Divisor = number of rows in all

Now, now... Play with me! C'mon, do math later!

2. [4] Bulbasaurs and [2] Charmeleons are taking a nap in the sunshine. How many legs are there in all?

Answer: _____

I made 5 columns.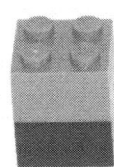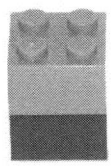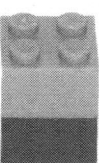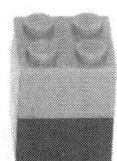

How many bricks are there per column?

I see 5 columns with 2 bricks per column:

$$10 \div 5 = 2$$

Dividend = number of bricks in all

Divisor = number of columns in all

Quotient = number of bricks per column

Division = Equal Sharing

You can find the number of bricks per column or you can find the number of bricks per row.

Yoo-Hoo! I got that all figured out, fellers! Hurry up! I'm bored! I want mazes!

1. I saw 9 Pokemon. Each third pokemon was a Squirtle. How many Squirtles did I see?

Answer: _____

1. <u>Write</u> the missing numbers, <u>multiply</u> and <u>divide</u>.

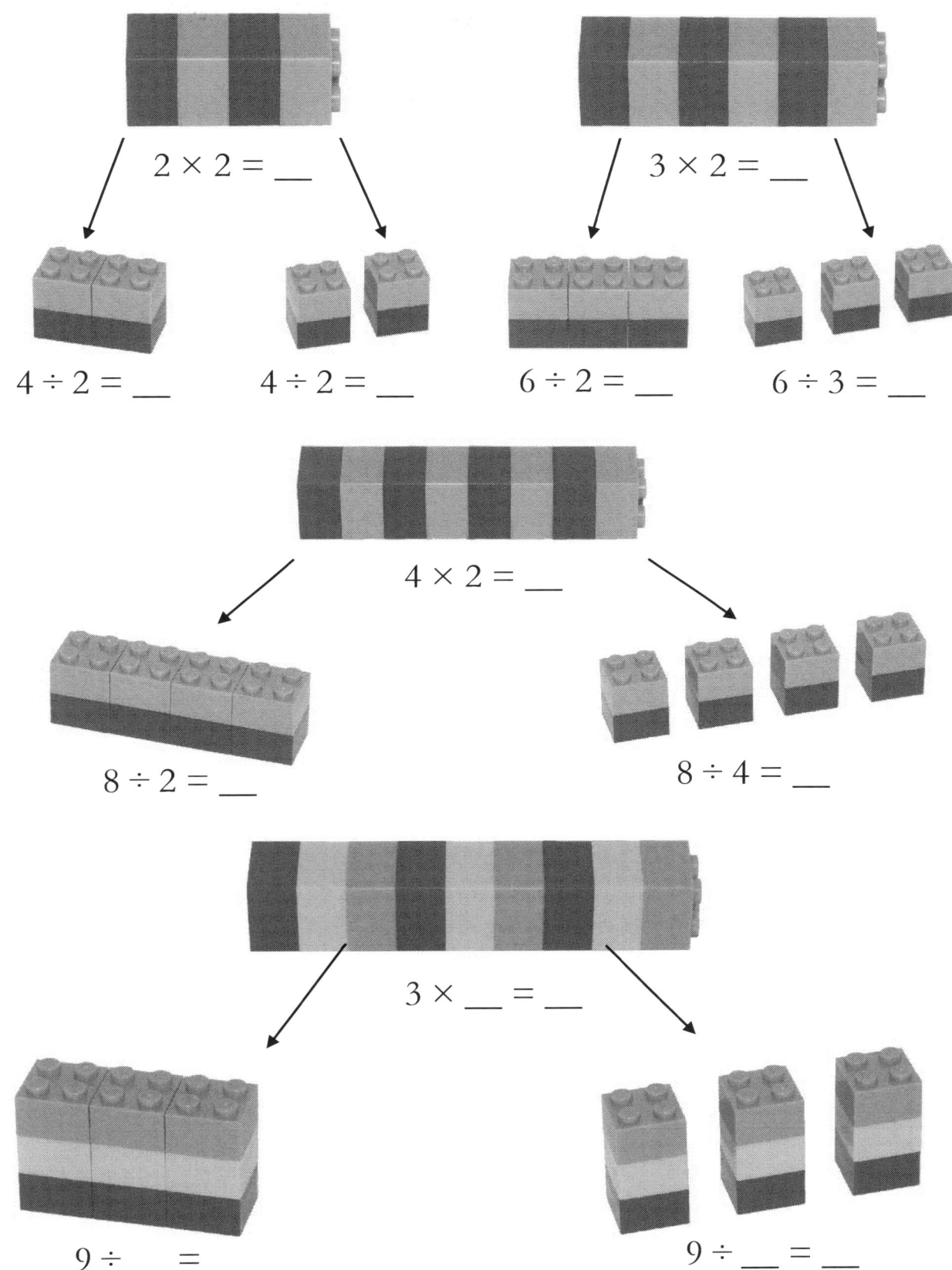

$2 \times 2 = \underline{}$

$4 \div 2 = \underline{}$ $4 \div 2 = \underline{}$

$3 \times 2 = \underline{}$

$6 \div 2 = \underline{}$ $6 \div 3 = \underline{}$

$4 \times 2 = \underline{}$

$8 \div 2 = \underline{}$ $8 \div 4 = \underline{}$

$3 \times \underline{} = \underline{}$

$9 \div \underline{} = \underline{}$ $9 \div \underline{} = \underline{}$

1. Write the missing numbers, multiply and divide.

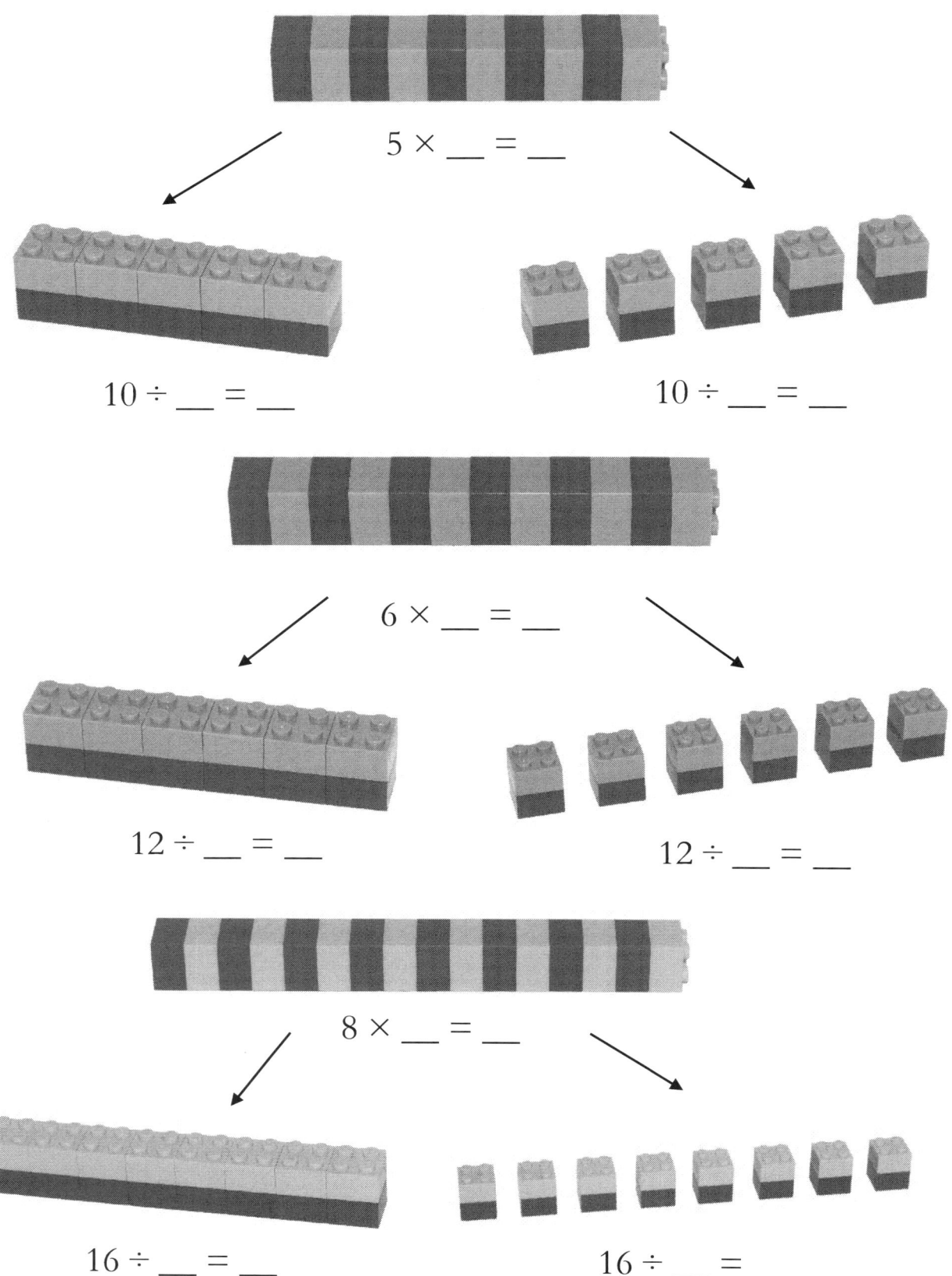

$5 \times \underline{} = \underline{}$

$10 \div \underline{} = \underline{}$ $10 \div \underline{} = \underline{}$

$6 \times \underline{} = \underline{}$

$12 \div \underline{} = \underline{}$ $12 \div \underline{} = \underline{}$

$8 \times \underline{} = \underline{}$

$16 \div \underline{} = \underline{}$ $16 \div \underline{} = \underline{}$

1. Write the missing numbers, multiply and divide.

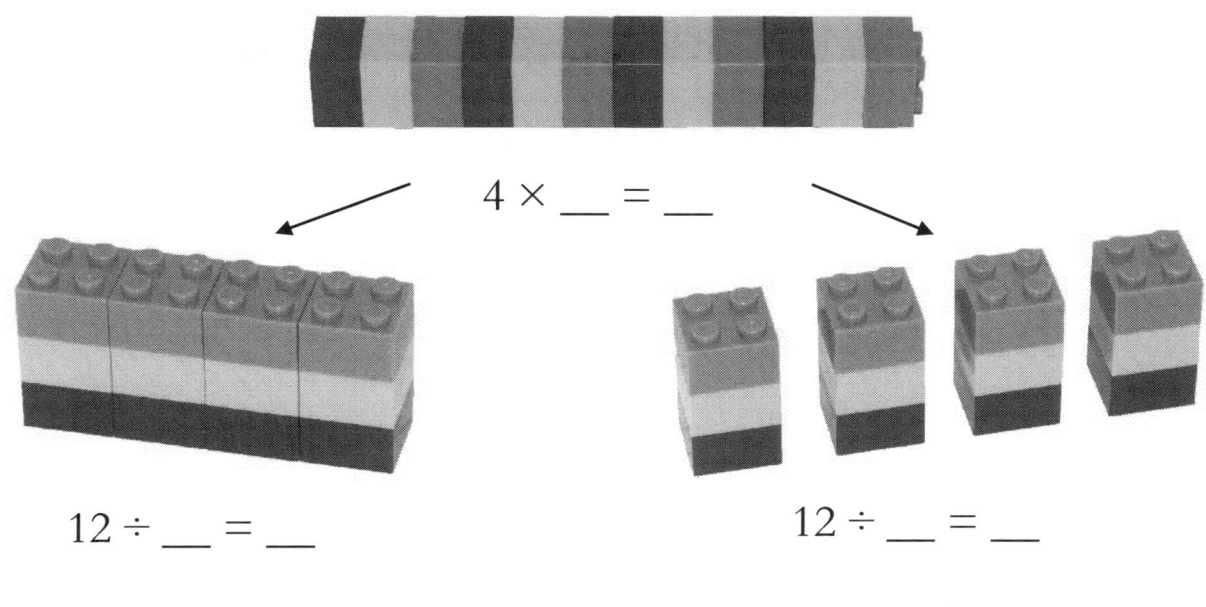

4 × __ = __

12 ÷ __ = __   12 ÷ __ = __

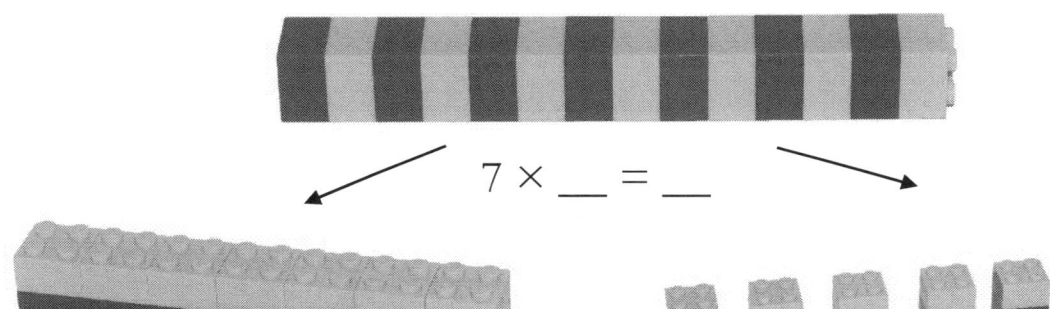

7 × __ = __

14  ÷ __ = __   14 ÷ __ = __

1. I saw 8 horns. If each Gogoat has 2 horns, <u>how many Gogoats</u> did I see?

Answer: _____

1. <u>Write</u> the missing numbers, <u>multiply</u> and <u>divide</u>.

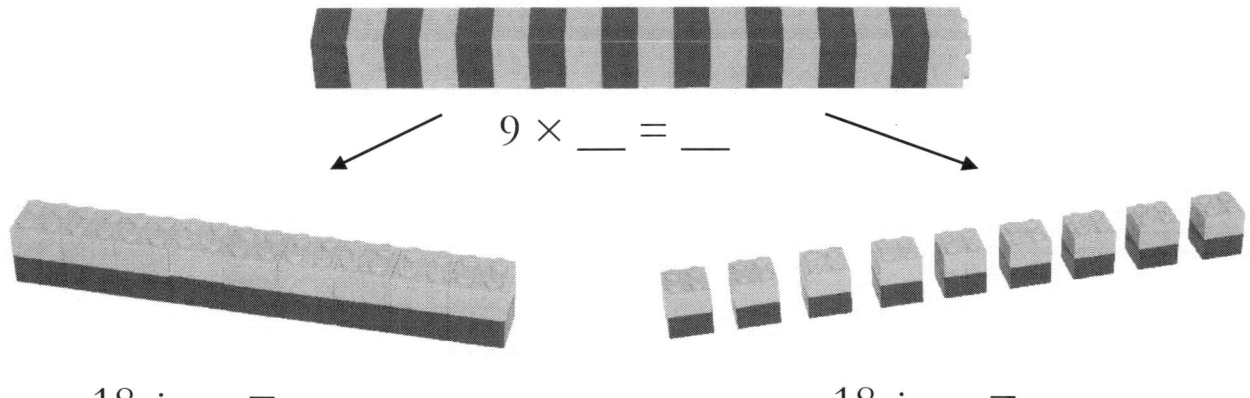

$9 \times \underline{} = \underline{}$

$18 \div \underline{} = \underline{}$ $\qquad$ $18 \div \underline{} = \underline{}$

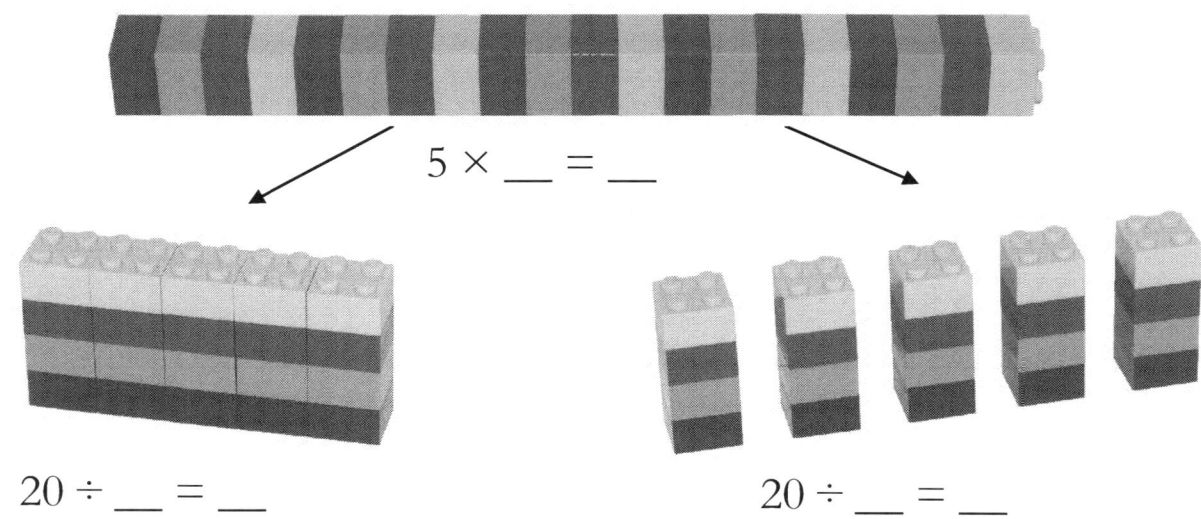

$5 \times \underline{} = \underline{}$

$20 \div \underline{} = \underline{}$ $\qquad$ $20 \div \underline{} = \underline{}$

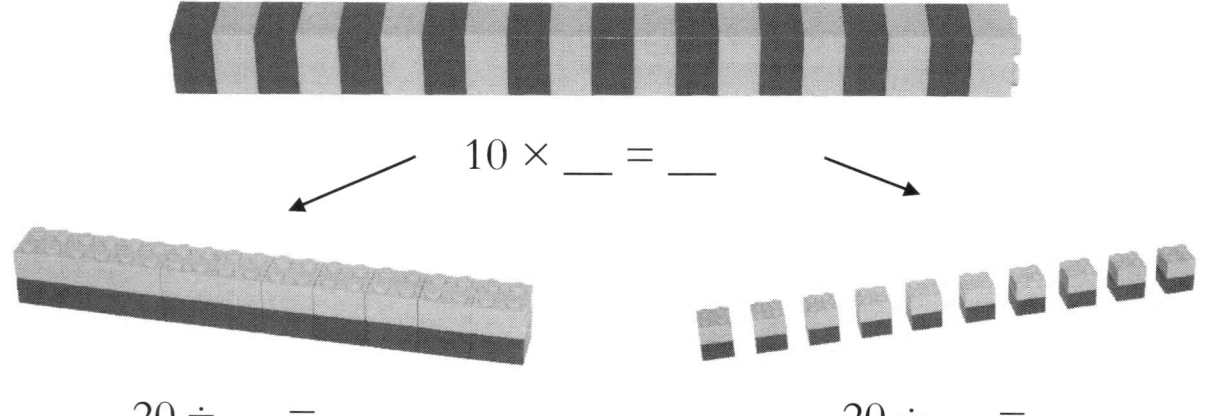

$10 \times \underline{} = \underline{}$

$20 \div \underline{} = \underline{}$ $\qquad$ $20 \div \underline{} = \underline{}$

Pokemon Coloring Math Book Multiplication and Division Part 1 Grades 1-4 Ages 6+

1. <u>Get</u> Pikachu.

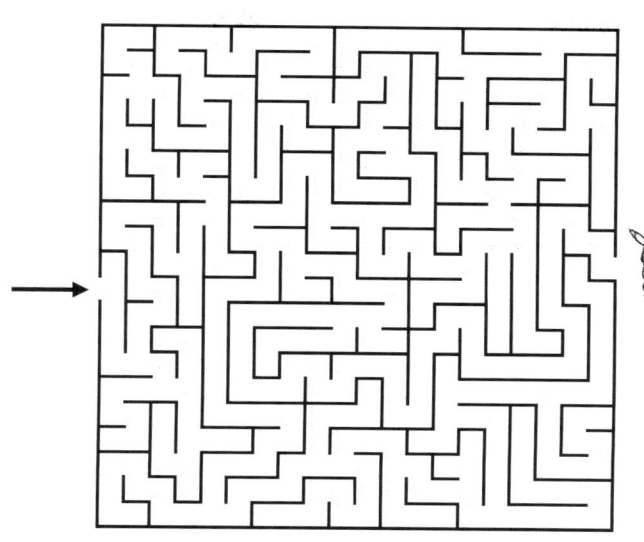

WELL-There sure are lots of paths in this tricky maze! Would you please solve it to me?

2. <u>Find</u> and <u>circle</u> or <u>cross out</u> the words.

```
E G E C H P I S X T Q E
O L Z M I O A K A P K T
P R E W E R O I M I R C
P V G C P R L V L V E N
O F Y A T E T E E J X I
N U L X C R S X L S L T
E R I F I R I M E G L X
N Y Z T O A V C C S R E
T J C H O R S E B A C K
N R O H Y H R I P I H W
P O N Y T A Z I Z M W P
X U C V P R D T E A J B
K A I I F A A N S S U M
E P T D S T D O F V D Z
Q Z Z H D M P T Y X W N
G I F C V Y J S X C C K
```

RHYHORN
FURY
ELECTRIC
TAIL
EXTREME
LAPRAS
EXTINCT
HORSELIKE
PONYTA
HORSEBACK
HOOVES
RAPIDASH
WHIP

1. <u>Write</u> the missing numbers, <u>multiply</u> and <u>divide</u>.

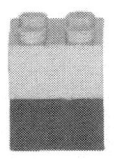  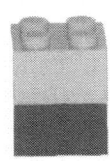

$2 \times 3 = \underline{}$

$6 \div 2 = \underline{}$

$6 \div 3 = \underline{}$

$2 \times 8 = \underline{}$

$\underline{} \div \underline{} = \underline{}$

$\underline{} \div \underline{} = \underline{}$

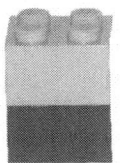 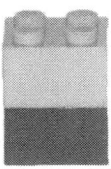 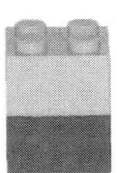 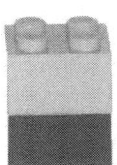 

$2 \times 5 = \underline{}$

$\underline{} \div \underline{} = \underline{}$

$\underline{} \div \underline{} = \underline{}$

$2 \times 9 = \underline{}$

$\underline{} \div \underline{} = \underline{}$

$\underline{} \div \underline{} = \underline{}$

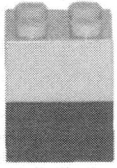 

$2 \times 2 = \underline{}$

$\underline{} \div \underline{} = \underline{}$

$\underline{} \div \underline{} = \underline{}$

2. I hatched <u>12</u> Pokemon. <u>Each fourth</u> Pokemon was Machop. <u>How many Machops</u> did I hatch?

Answer: _____

1. <u>Write</u> the missing numbers, <u>multiply</u> and <u>divide</u>.

"The easiest way to get math done is to have fun doing it!"

2 × 4 = __

8 ÷ 2 = __

8 ÷ 4 = __

2 × 7 = __

__ ÷ __ = __

__ ÷ __ = __

2 × 10 = __

__ ÷ __ = __

__ ÷ __ = __

2. Slowpoke spent ⟦twice⟧ as much time along the riverbank as Slowbro. Slowbro spent ⟦6⟧ minutes. <u>How much time</u> did Slowpoke spend along the riverbank?

2 × 6 = __

__ ÷ __ = __

__ ÷ __ = __

Answer: _____

1. <u>Multiply</u>.

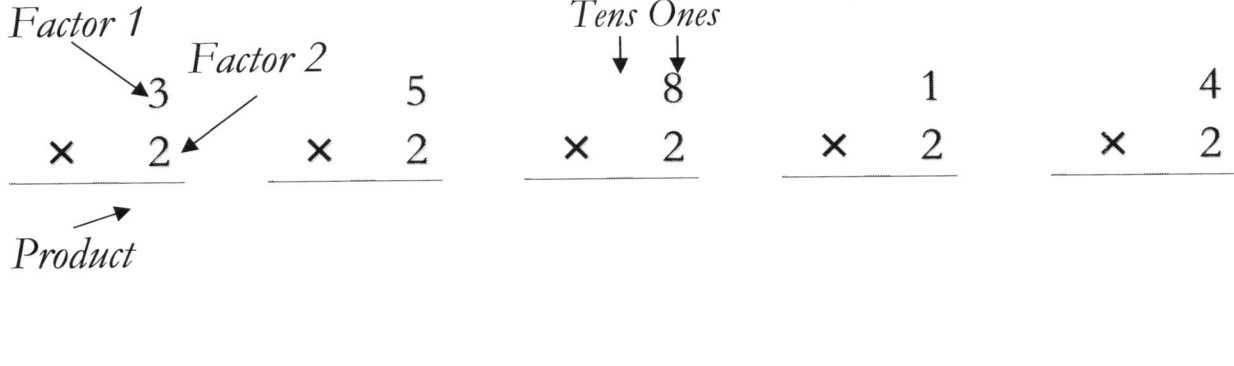

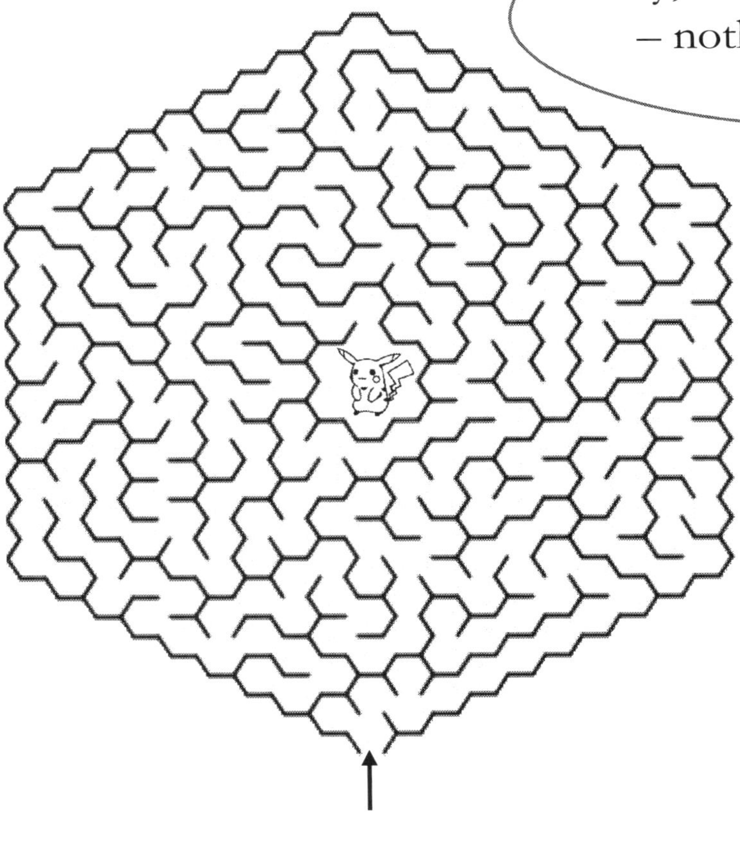

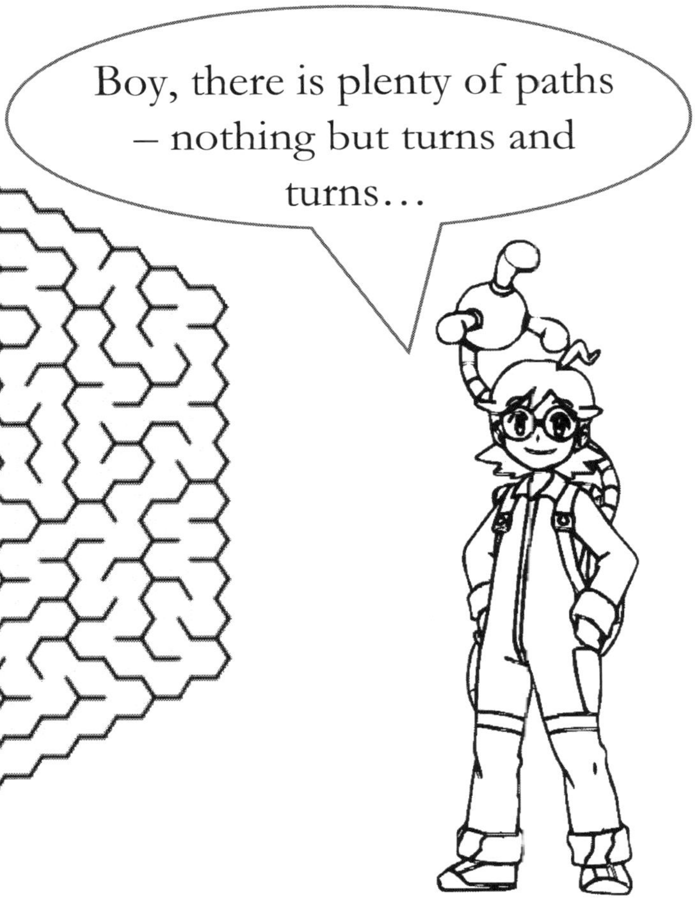

1. Make 2 rows. X shows the number of columns. Find X. Write the missing numbers, multiply and divide.

2 × X = 10   X = __
10 ÷ 5 = __   10 ÷ 2 = __

__ rows

__ bricks in each row

2 × X = 16   X = __
16 ÷ 8 = __   16 ÷ 2 = __

__ rows   __ bricks in each row

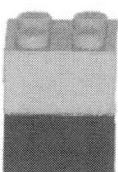 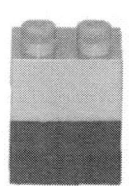 

2 × X = 6   X = __
6 ÷ 3 = __   6 ÷ 2 = __

__ rows

__ bricks in each row

2. Wartortle gets 2 scratches for each battle. It had 7 battles. How many scratches does it have in all?

Answer: _____

1. Make 2 rows. X shows the number of columns. Find X. Write the missing numbers, multiply and divide.

         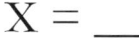

$2 \times X = 20$ $X = \underline{}$

$20 \div 10 = \underline{}$ $20 \div 2 = \underline{}$

__ rows   __ bricks in each row

$2 \times X = 8$ $X = \underline{}$

$8 \div 4 = \underline{}$ $8 \div 2 = \underline{}$

__ rows

__ bricks in each row

      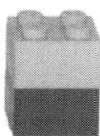

$2 \times X = 14$ $X = \underline{}$

$14 \div 7 = \underline{}$ $14 \div 2 = \underline{}$

__ rows   __ bricks in each row

2. Blastoise fired 16 water bullets. Each second bullet hit the target. How many water bullets hit the target?

Answer: _____

1. Make 2 rows. X shows the number of columns. Find X. Write the missing numbers, multiply and divide.

 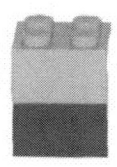  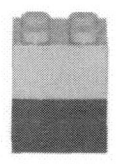  

2 × X = 12   X = __

12 ÷ 6 = __   12 ÷ 2 = __

__ rows   __ bricks in each row

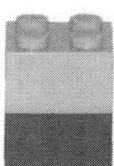 

2 × X = 4   X = __

4 ÷ 2 = __   4 ÷ 2 = __

__ rows

__ bricks in each row

2 × X = 18   X = __

18 ÷ 9 = __ 18 ÷ 2 = __

__ rows   __ bricks in each row

2. Mankey had 5 fits of temper. Each fit was preceded by 3 violent tremors. How many tremors did it have in all?

Answer: _____

1. <u>Write</u> the missing numbers.

× 2 / 8   × 2 / 4   × 2 / 20   × 2 / 12   × 2 / 18

× 2 / 10   × 2 / 16   × 2 / 6   × 2 / 14   × 2 / 2

Perhaps this time I will get this little monster!

1. <u>Add</u> and <u>multiply</u>.

2 + 2 + 2 = __   3 × 2 = __

2 + 2 + 2 + 2 = __   4 × 2 = __

2 + 2 + 2 + 2 + 2 = __   5 × 2 = __

2 + 2 + 2 + 2 + 2 + 2 = __   6 × 2 = __

2 + 2 + 2 + 2 + 2 + 2 + 2 = __   7 × 2 = __

2 + 2 + 2 + 2 + 2 + 2 + 2 + 2 = __   8 × 2 = __

2 + 2 + 2 + 2 + 2 + 2 + 2 + 2 + 2 = __   9 × 2 = __

2 + 2 + 2 + 2 + 2 + 2 + 2 + 2 + 2 + 2 = __   10 × 2 = __

2. <u>Take</u> __ __ bricks to answer the questions.

<u>Make</u> 2 rows.

<u>How many</u> columns can you make with 12 bricks?   __

<u>How many</u> columns can you make with the leftover bricks?   __

<u>How many</u> columns can you make out of all the bricks?   __ .

1. <u>Get</u> Pikachu.

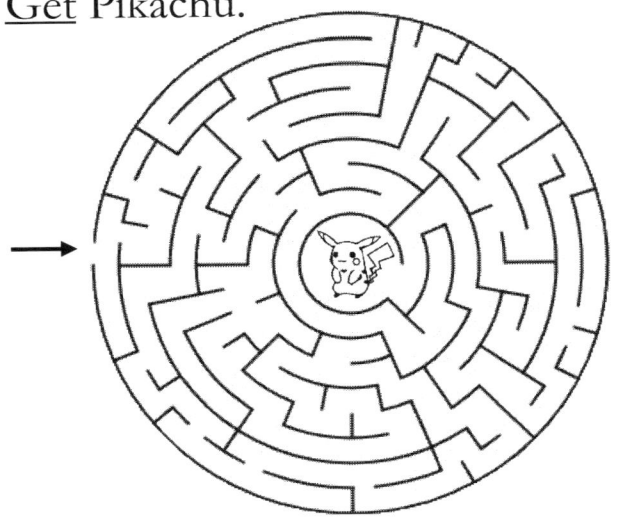

2. <u>Find</u> and <u>circle</u> or <u>cross out</u> the words.

> This word search looks like an adventure! It has adjectives!
> W-O-O-S-H!

```
V C A F C A A C T W G M
T R Z A E L F L N H N D
T D F S K R E E C V I G
I E R T A J R F D Z R I
A O L G M A O A A X A L
H Y I T R B A I O B C O
S L A E N B S R O T L H
E X L V G E A Y E X X E
J Z P X G W G P X R L D
O J M R W U A K M B T D
T E T K F Z Z T C N I L
J E C A F R U S E C Q M
L J Q N N V J X U R Y N
F Y S X O S P E E D U U
E T E C H N I Q U E S M
B Y X U R N X U C J O S
```

CLEFABLE

CLEFAIRY

GENTLE

CARING

TECHNIQUES

FAST

SPEED

RARE

HORSEA

FRAGILE

SURFACE

WATER

1. <u>Write</u> the missing numbers, <u>multiply</u> and <u>divide</u>.

$3 \times 3 =$ ___

$9 \div 3 =$ ___

$9 \div 3 =$ ___

$3 \times 2 =$ ___

___ $\div$ ___ $=$ ___

___ $\div$ ___ $=$ ___

$3 \times 8 =$ ___

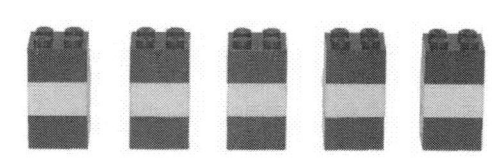

___ $\div$ ___ $=$ ___

___ $\div$ ___ $=$ ___

$3 \times 5 =$ ___

___ $\div$ ___ $=$ ___

___ $\div$ ___ $=$ ___

$3 \times 9 =$ ___

___ $\div$ ___ $=$ ___

___ $\div$ ___ $=$ ___

2. Rhyhorn thinks about demolishing each of its target ⬚3⬚ times a day. If it has ⬚4⬚ targets, <u>how many thoughts</u> does it have daily?

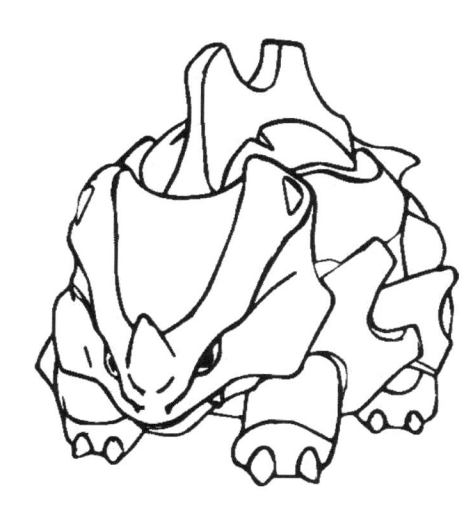

Answer: _____

1. <u>Write</u> the missing numbers, <u>multiply</u> and <u>divide</u>.

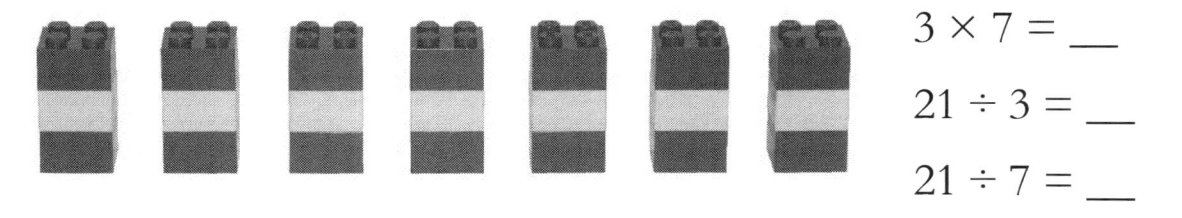

$3 \times 7 = \underline{}$

$21 \div 3 = \underline{}$

$21 \div 7 = \underline{}$

$3 \times 4 = \underline{}$

$\underline{} \div \underline{} = \underline{}$

$\underline{} \div \underline{} = \underline{}$

$3 \times 10 = \underline{}$ $\underline{} \div \underline{} = \underline{}$ $\underline{} \div \underline{} = \underline{}$

$3 \times 6 = \underline{}$

$\underline{} \div \underline{} = \underline{}$

$\underline{} \div \underline{} = \underline{}$

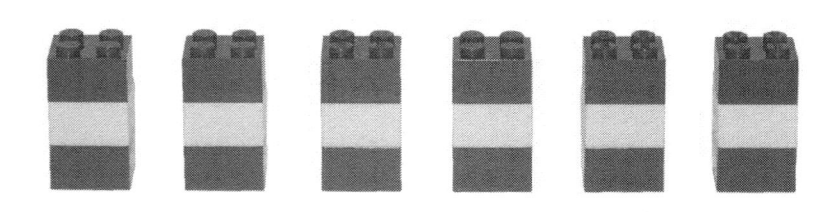

2. Jolteon had ３ battles. Each battle it called down ４ thunderbolts! <u>How many thunderbolts</u> did it call down in all?

Answer: _____

1. <u>Multiply</u>.

$$\begin{array}{r}3\\ \times\ 3\\ \hline\end{array} \qquad \begin{array}{r}5\\ \times\ 3\\ \hline\end{array} \qquad \begin{array}{r}8\\ \times\ 3\\ \hline\end{array} \qquad \begin{array}{r}1\\ \times\ 3\\ \hline\end{array} \qquad \begin{array}{r}4\\ \times\ 3\\ \hline\end{array}$$

$$\begin{array}{r}6\\ \times\ 3\\ \hline\end{array} \qquad \begin{array}{r}9\\ \times\ 3\\ \hline\end{array} \qquad \begin{array}{r}7\\ \times\ 3\\ \hline\end{array} \qquad \begin{array}{r}2\\ \times\ 3\\ \hline\end{array} \qquad \begin{array}{r}10\\ \times\ 3\\ \hline\end{array}$$

Look! There Pikachu is! What are you waiting for? GET IT!

1. Make ⒊ rows. ⓧ shows the number of columns. Find X. Write the missing numbers, multiply and divide.

$3 \times X = 30$ $X = \_\_$

$30 \div 10 = \_\_$ $30 \div 3 = \_\_$

__ rows   __ bricks in each row

$3 \times X = 18$ $X = \_\_$

$18 \div 6 = \_\_$ $18 \div 3 = \_\_$

__ rows

__ bricks in each row

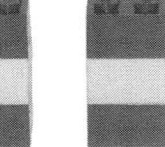

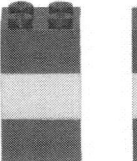

$3 \times X = 27$ $X = \_\_$

$27 \div 9 = \_\_$ $27 \div 3 = \_\_$

__ rows   __ bricks in each row

2. Moltres found ⒊ volcanoes. Each time he bathed ⒌ times in lava. How many times did he bathe in all?

Answer: _____

1. Make 3 rows. X shows the number of columns. Find X. Write the missing numbers, multiply and divide.

$3 \times X = 15$ $X = \_\_$
$15 \div 5 = \_\_$ $15 \div 3 = \_\_$
__ rows

__ bricks in each row

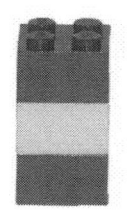

$3 \times X = 24$ $X = \_\_$
$24 \div 8 = \_\_$ $24 \div 3 = \_\_$
__ rows

__ bricks in each row

The Charmander egg will be hatched any minute! Speed up! I've got to get math done!

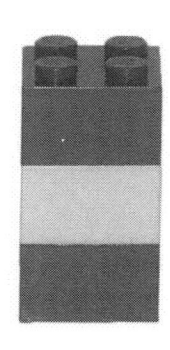

$3 \times X = 6$ $X = \_\_$
$6 \div 2 = \_\_$ $6 \div 3 = \_\_$
__ rows
__ bricks in each row

1. <u>Make</u> 3 rows. X shows the number of columns. <u>Find</u> X. <u>Write</u> the missing numbers, <u>multiply</u> and <u>divide</u>.

$3 \times X = 21$ X = __
$21 \div 7 =$ __ $21 \div 3 =$ __
__ rows
__ bricks in each row

$3 \times X = 9$ X = __
$9 \div 3 =$ __ $9 \div 3 =$ __
__ rows   __ bricks in each row

$3 \times X = 12$ X = __
$12 \div 4 =$ __ $12 \div 3 =$ __
__ rows
__ bricks in each row

2. Moltres bathed 24 times in lava. If he bathed 8 times in each volcano, <u>how many volcanoes</u> did he dive into?

Answer: _____

1. <u>Write</u> the missing numbers.

× 3	× 3	× 3	× 3	× 3
9	3	21	12	18

× 3	× 3	× 3	× 3	× 3
15	27	6	24	30

Now, with such a Smarty and such a Sunny, how can we fail?

1. <u>Add</u> and <u>multiply</u>. The easiest way to count 3's is counting by groups of two 3's or of four 3's. Two 3's (3 + 3) equal 6. Or four 3's (3 + 3 + 3 + 3) equal 12.

3 + 3 = __   2 × 3 = __

3 + 3 + 3 = 6 + 3 = __   3 × 3 = __

3 + 3 + 3 + 3 = __   4 × 3 = __

3 + 3 + 3 + 3 + 3 = __   5 × 3 = __

3 + 3 + 3 + 3 + 3 + 3 = __   6 × 3 = __

3 + 3 + 3 + 3 + 3 + 3 + 3 = __   7 × 3 = __

3 + 3 + 3 + 3 + 3 + 3 + 3 + 3 = __   8 × 3 = __

3 + 3 + 3 + 3 + 3 + 3 + 3 + 3 + 3 = __   9 × 3 = __

3 + 3 + 3 + 3 + 3 + 3 + 3 + 3 + 3 + 3 = __   10 × 3 = __

2. <u>Take</u> __ __ bricks to answer the questions.

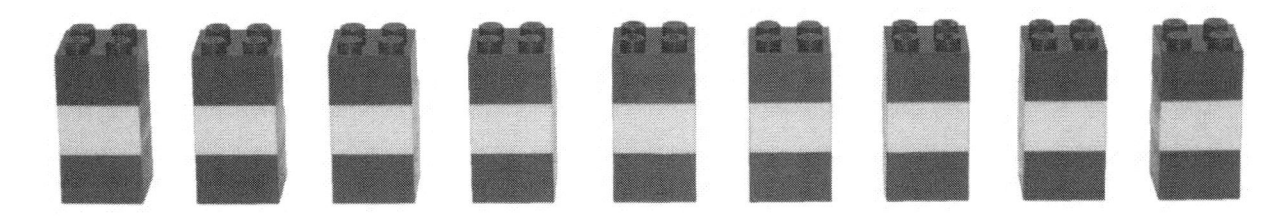

<u>Make</u> 3 rows.

<u>How many</u> columns can you make with 15 bricks?   __

<u>How many</u> columns can you make with the leftover bricks?   __

<u>How many</u> columns can you make out of all the bricks?   __.

1. <u>Get</u> Pikachu.

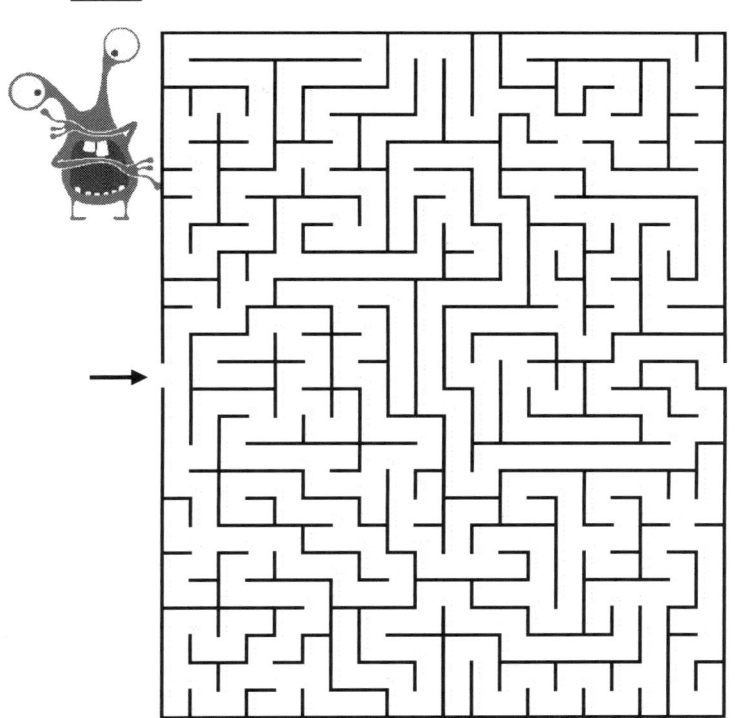

Pickles, you're making me nervous! I wish I'd be hatched into something BIG!

2. <u>Find</u> and <u>circle</u> or <u>cross out</u> the words.

U	P	E	Q	T	O	M	W	P	H	W	H	T
T	N	O	A	O	N	O	N	E	Y	Y	I	R
H	N	D	V	T	R	A	L	C	P	X	P	E
S	X	Y	E	R	E	J	D	N	N	I	S	S
W	H	O	U	R	B	R	O	N	O	Q	A	E
E	J	B	N	C	G	T	L	U	E	T	N	D
P	I	C	K	Y	I	R	F	A	R	P	D	N
R	O	N	T	Z	T	C	O	W	F	E	S	J
S	Z	G	E	J	O	X	C	U	E	F	H	G
G	N	I	T	H	G	I	F	P	N	Q	R	D
S	C	M	R	B	E	W	T	J	E	D	E	I
E	A	T	N	D	O	X	X	T	W	X	W	G
V	K	F	T	F	U	U	E	A	Q	O	D	K

SANDSHREW
DESERT
UNDERGROUND
FIGHTING
PICKY
BURROW
EATER
DEEP
HYPNO
PENDANT
HYPNOTIZE

1. <u>Multiply</u> and <u>divide</u>.

2 × 4 = __	2 × ❀ = 8	❀ = __	8 ÷ 2 = __
2 × 2 = __	2 × ❀ = 4	❀ = __	4 ÷ 2 = __
2 × 5 = __	2 × ❀ = 10	❀ = __	10 ÷ 2 = __
2 × 3 = __	2 × ❀ = 6	❀ = __	6 ÷ 2 = __
3 × 3 = __	3 × ❀ = 9	❀ = __	9 ÷ 3 = __
3 × 2 = __	3 × ❀ = 6	❀ = __	6 ÷ 3 = __
4 × 2 = __	4 × ❀ = 8	❀ = __	8 ÷ 4 = __
5 × 2 = __	5 × ❀ = 10	❀ = __	10 ÷ 5 = __

2. <u>Divide</u> (<u>find how many groups of 2, 3, 4, or 5</u> you can make out of all the bricks) and <u>write down</u> the remainder in the parentheses.

Remainder is what is left over after the division operation.

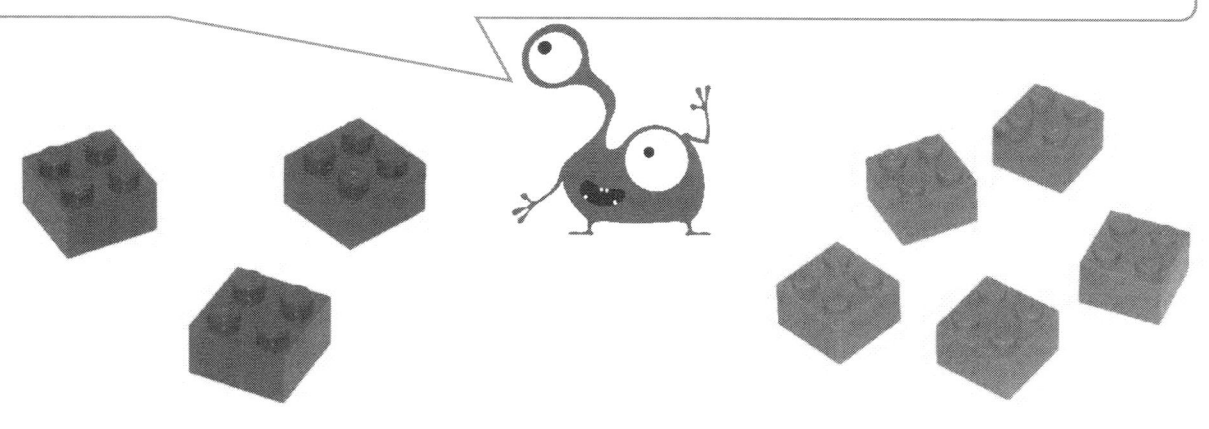

3 ÷ 2 = 1 (1)

5 ÷ 2 = __ (__)

5 ÷ 5 = __ (__)

3 ÷ 3 = __ (__)

5 ÷ 4 = __ (__)

5 ÷ 3 = __ (__)

1. <u>Divide</u> and <u>write</u> the remainder in the parentheses.

$6 \div 5 = 1 \ (1)$

$6 \div 3 = \underline{\ \ } \ (\underline{\ \ })$

$6 \div 6 = \underline{\ \ } \ (\underline{\ \ })$

$6 \div 2 = \underline{\ \ } \ (\underline{\ \ })$

$6 \div 4 = \underline{\ \ } \ (\underline{\ \ })$

$4 \div 2 = \underline{\ \ } \ (\underline{\ \ })$

$4 \div 3 = \underline{\ \ } \ (\underline{\ \ })$

$4 \div 4 = \underline{\ \ } \ (\underline{\ \ })$

$7 \div 2 = 3 \ (1)$

$7 \div 5 = \underline{\ \ } \ (\underline{\ \ })$

$7 \div 6 = \underline{\ \ } \ (\underline{\ \ })$

$7 \div 4 = \underline{\ \ } \ (\underline{\ \ })$

$7 \div 3 = \underline{\ \ } \ (\underline{\ \ })$

$7 \div 7 = \underline{\ \ } \ (\underline{\ \ })$

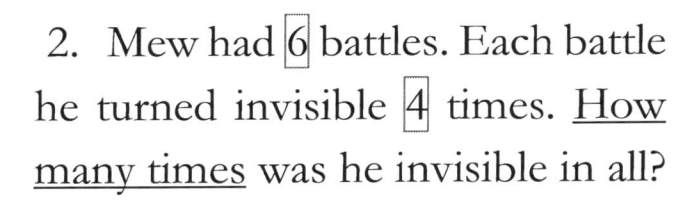

2. Mew had 6 battles. Each battle he turned invisible 4 times. <u>How many times</u> was he invisible in all?

Answer: _____.

1. <u>Multiply</u> and <u>divide</u>.

$8 \div 2 = 4$

$18 \div 2 = \underline{}$

$2 = \underline{}$

$6 = 12$

$8 = \underline{}$

$12 \div 2 = \underline{}$

$10 \div 2 = \underline{}$

$3 = \underline{}$

$2 \times$

$5 = \underline{}$

$14 \div 2 = \underline{}$

$16 \div 2 = \underline{}$

$7 = \underline{}$

$9 = \underline{}$

$6 \div 2 = \underline{}$

$4 \div 2 = \underline{}$

$4 = \underline{}$

2. <u>Find</u> the missing divisor.

$15 \div X = 3$ $18 \div X = 6$ $12 \div X = 4$

$X = \underline{}$ $X = \underline{}$ $X = \underline{}$

$9 \div ? = 3$ $8 \div ? = 2$ $6 \div ? = 2$

$? = \underline{}$ $? = \underline{}$ $? = \underline{}$

$12 \div 🐭 = 2$ $16 \div 🐻 = 4$ $20 \div 🐖 = 5$

$🐭 = \underline{}$ $🐻 = \underline{}$ $🐖 = \underline{}$

1. <u>Divide</u> and <u>write</u> the remainder in the parentheses.

$9 \div 5 = \underline{} (\underline{})$

$9 \div 2 = \underline{} (\underline{})$

$9 \div 7 = \underline{} (\underline{})$

$9 \div 4 = \underline{} (\underline{})$

$9 \div 6 = \underline{} (\underline{})$

$9 \div 3 = \underline{} (\underline{})$

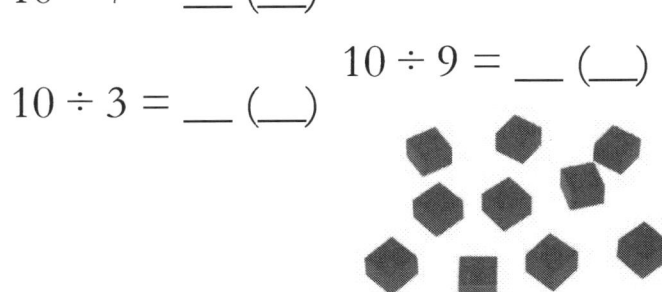

$8 \div 2 = 4 \ (0)$

$8 \div 8 = \underline{} (\underline{})$

$8 \div 6 = \underline{} (\underline{})$

$8 \div 5 = \underline{} (\underline{})$

$8 \div 3 = \underline{} (\underline{})$

$8 \div 4 = \underline{} (\underline{})$

$8 \div 7 = \underline{} (\underline{})$

$10 \div 6 = \underline{} (\underline{})$

$10 \div 5 = \underline{} (\underline{})$

$10 \div 8 = \underline{} (\underline{})$

$10 \div 2 = \underline{} (\underline{})$

$10 \div 4 = \underline{} (\underline{})$

$10 \div 7 = \underline{} (\underline{})$

$10 \div 9 = \underline{} (\underline{})$

$10 \div 3 = \underline{} (\underline{})$

1. <u>Write</u> the missing numbers $\boxed{2, 2, 2, 5, \text{ and } 10}$ to make the equation true.

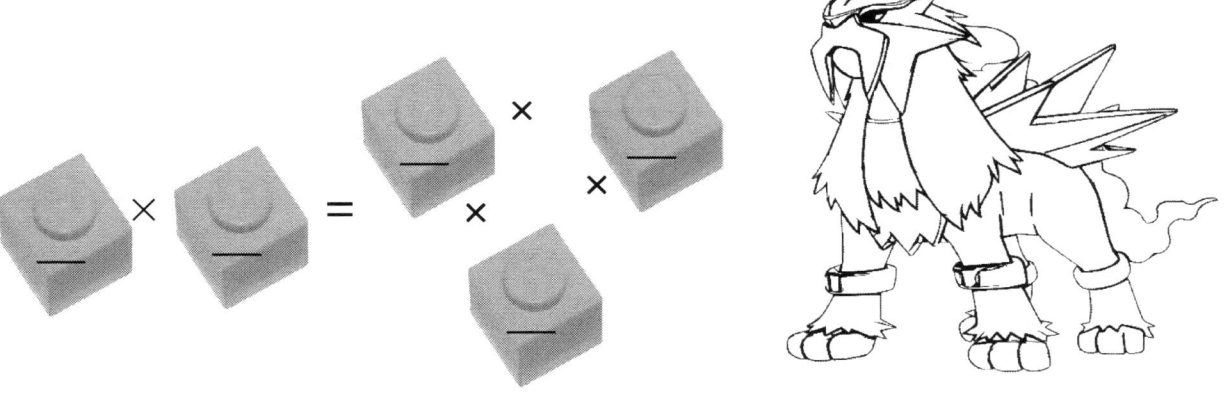

☐ × ☐ = ☐ × ☐ × ☐

2. <u>Find</u> the missing factor or dividend.

$X \times 2 = 18$	$X \times 7 = 14$	$X \times 4 = 12$
$X = \_\_$	$X = \_\_$	$X = \_\_$
$? \times 5 = 15$	$? \times 5 = 20$	$? \times 2 = 12$
$? = \_\_$	$? = \_\_$	$? = \_\_$
🐭 $\times 6 = 18$	🐰 $\times 4 = 16$	🐖 $\times 2 = 16$
🐭 $= \_\_$	🐰 $= \_\_$	🐖 $= \_\_$
$X \div 3 = 4$	$X \div 3 = 6$	$X \div 9 = 2$
$X = \_\_$	$X = \_\_$	$X = \_\_$
$? \div 6 = 2$	$? \div 4 = 5$	$? \div 3 = 5$
$? = \_\_$	$? = \_\_$	$? = \_\_$
🐭 $\div 10 = 2$	🐰 $\div 8 = 1$	🐖 $\div 1 = 4$
🐭 $= \_\_$	🐰 $= \_\_$	🐖 $= \_\_$

1. <u>Get</u> Pikachu.

   Here's where I get even for all the Team Rocket that ever pestered me.

2. <u>Find</u> and <u>circle</u> or <u>cross out</u> the words.

```
S L E N N U T T H U L E
E H N I V E A O L C Z S
I C C K Y B T T M A C U
I L G E L Z R I Z V J F
P K A O E A U R B E A N
V H G T S R A B S S Z O
G T K O H D C N A X M C
Q B N S A W G S B T I M
N I H R S G N A F Y V Y
C X T W Q K P H U C K R
N I M J A M I E U W N R
E N E R G Y E C Y N K J
```

ZUBAT
TUNNELS
CAVES
ULTRASONIC
RADAR
GOLBAT
SCREECH
FANGS
BITE
CONFUSE
ENERGY

1. <u>Write</u> the missing numbers $\boxed{2, 2, 2, 2, 3, 3, \text{ and } 4}$ to make the equation true.

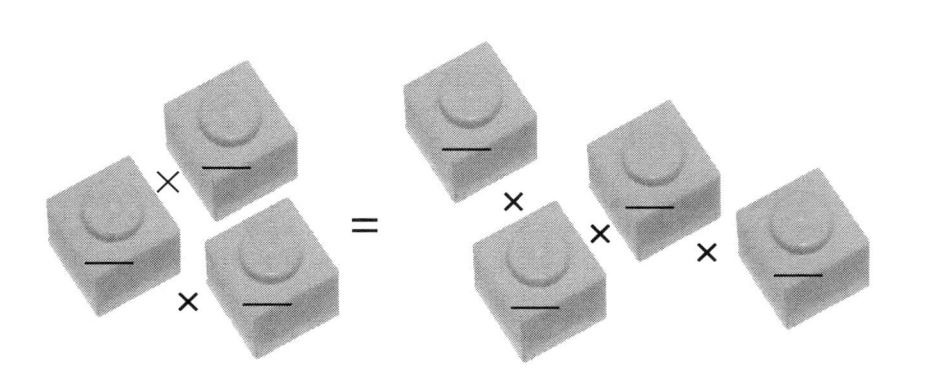

___ × ___ × ___ = ___ × ___ × ___ × ___

2. <u>Find</u> the missing factor or dividend.

X × 3 = 12	X × 3 = 18	X × 9 = 18
X = __	X = __	X = __
? × 6 = 12	? × 4 = 20	? × 3 = 15
? = __	? = __	? = __
🐭 × 10 = 20	🐰 × 8 = 8	🐷 × 1 = 4
🐭 = __	🐰 = __	🐷 = __
X ÷ 2 = 9	X ÷ 7 = 2	X ÷ 4 = 3
X = __	X = __	X = __
? ÷ 5 = 3	? ÷ 5 = 4	? ÷ 2 = 6
? = __	? = __	? = __
🐭 ÷ 6 = 3	🐰 ÷ 4 = 4	🐷 ÷ 2 = 8
🐭 = __	🐰 = __	🐷 = __

1. <u>Compare</u> the number sentences, using ">", "<", or "=".

4 + 4 + 4 __ 5 × 4		2 + 2 + 2 + 2 __ 2 × 2	
3 + 3 __ 4 × 3		5 + 5 + 5 __ 3 × 5	
6 + 6 __ 3 × 6		2 + 2 + 2 __ 4 × 2	
3 + 3 + 3 + 3 __ 5 × 3		7 + 7 __ 2 × 7	
8 + 8 __ 1 × 8		3 + 3 + 3 __ 2 × 3	

2. <u>Evaluate</u> each expression. <u>Indicate</u> the order of operations.

Are you kidding? First, we multiply in the parentheses (3 × 3 = 9), then, we add or subtract: 9 + 15 = 24.

3 × 5 = __	(3 × 3) $\overset{1}{+}$ $\overset{2}{15}$ = __	(3 × 6) $\overset{1}{-}$ $\overset{2}{12}$ = __
3 × 2 = __	(3 × 4) + 8 = __	(3 × 10) − 15 = __
3 × 6 = __	(3 × 7) + 19 = __	(3 × 4) − 6 = __
3 × 9 = __	(3 × 8) + 7 = __	(3 × 5) − 7 = __

2. I captured ⎡3⎤ Pidgeys om Monday. I captured ⎡four times⎤ as many Pidgeys on Friday as on Monday. <u>How many Pidgeys</u> did I capture in all?

Answer: _____.

1. <u>Write</u> the missing numbers $\boxed{2, 2, 2, 2, 4, 4, \text{and } 4}$ to make the equation true.

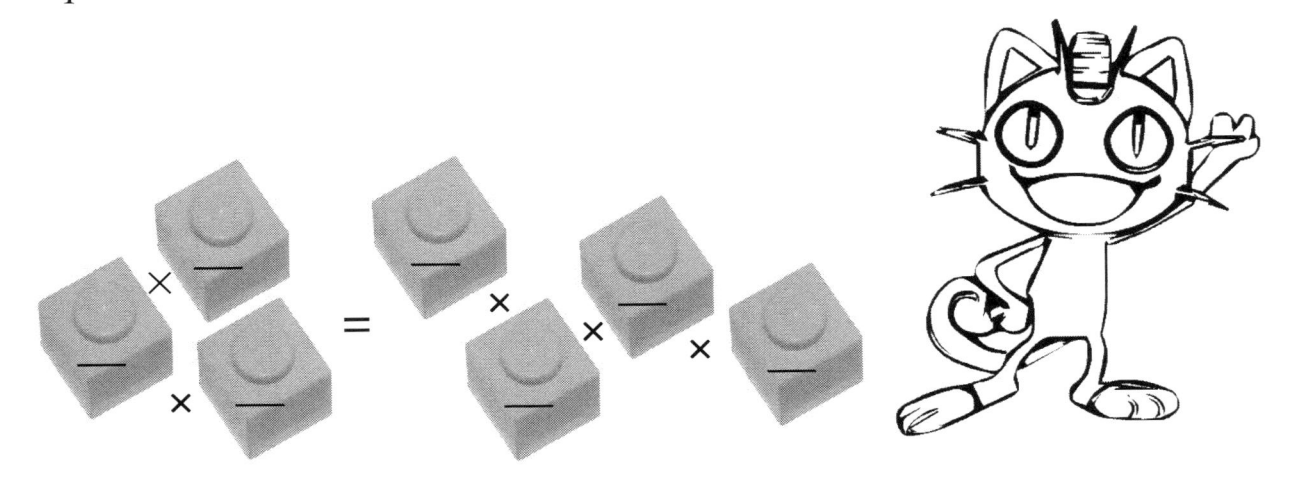

2. <u>Evaluate</u> each expression. <u>Indicate</u> the order of operations.

$$\overset{1}{(6 \times 2)} \overset{2}{- 5} \overset{3}{+ 10} \overset{4}{- 8} = 12 - 5 + 10 - 8 = 7 + 10 - 8 = 17 - 8 = 9$$

$(4 \times 4) - 3 + 5 - 9 = $ _____

$(3 \times 5) - 8 + 4 - 6 = $ _____

$(5 \times 4) - 6 + 3 - 8 = $ _____

$(5 \times 2) - 4 + 12 - 3 = $ _____

$(4 \times 2) - 1 + 13 - 6 = $ _____

$(3 \times 6) - 13 + 7 - 4 = $ _____

$(3 \times 4) - 6 + 11 - 4 = $ _____

1. <u>Multiply</u> and <u>divide</u>. <u>Connect</u> the three multiplication and division diagrams of the same value. The first one is done for you.

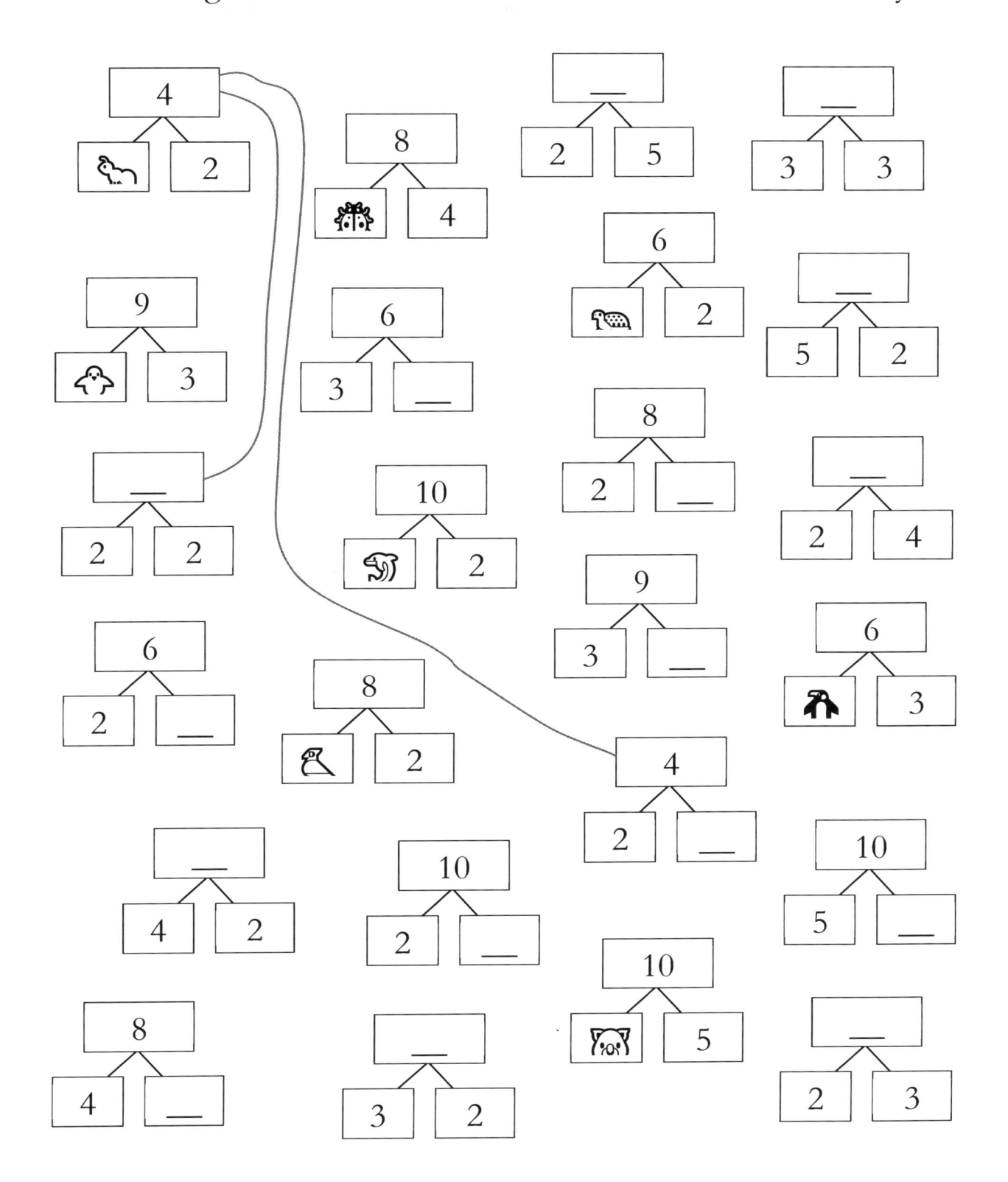

1. <u>Write</u> the missing numbers ⎡2, 2, 3, 3, 3, 3, and 4⎤ to make the equation true.

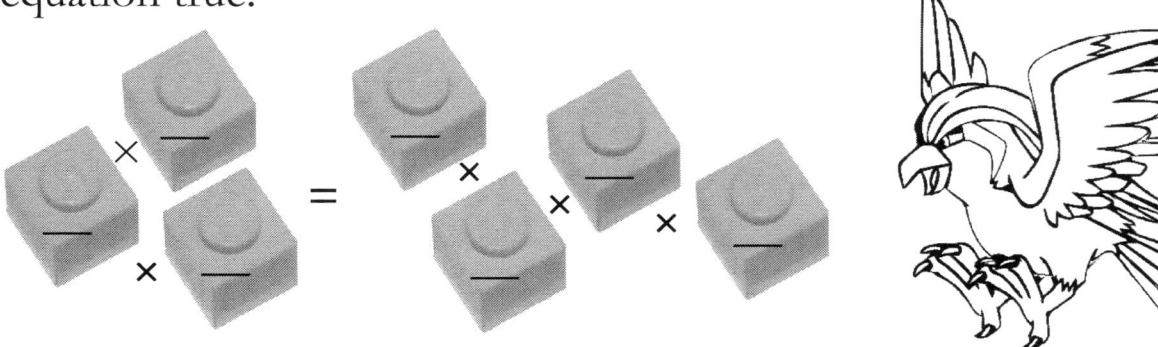

___ × ___ × ___ = ___ × ___ × ___ × ___

2. <u>Find</u> the missing divisor.

$18 \div X = 2$ $15 \div X = 5$ $18 \div X = 3$

X = ___   X = ___   X = ___

$20 \div ? = 4$ $16 \div ? = 8$ $8 \div ? = 4$

? = ___   ? = ___   ? = ___

$18 \div 🐿 = 9$ $16 \div 🐇 = 2$ $14 \div 🐖 = 7$

🐿 = ___   🐇 = ___   🐖 = ___

3. <u>Evaluate</u> each expression. <u>Indicate</u> the order of operations.

$(3 \times 6) - 15 + 17 - 4 =$ ___ $(3 \times 4) - 11 + 19 - 8 =$ ___

$(7 \times 2) - 12 + 14 - 8 =$ ___ $(2 \times 9) - 17 + 14 - 7 =$ ___

$(6 \times 3) - 13 + 10 - 6 =$ ___ $(2 \times 8) - 12 + 9 - 11 =$ ___

4. Pidgeotto used ⎡6⎤ times as many claws as pecks on his intruder. If it fiercely used ⎡5⎤ peckes on his intruder, <u>how many times</u> did it use its claws?

Answer: _____.

1. <u>Multiply</u> and <u>divide</u>.

$2 \times 8 = \_\_$	$2 \times ❄ = 16$	$❄ = \_\_$	$16 \div 2 = \_\_$
$2 \times 6 = \_\_$	$2 \times ❄ = 12$	$❄ = \_\_$	$12 \div 2 = \_\_$
$2 \times 10 = \_\_$	$2 \times ❄ = 20$	$❄ = \_\_$	$20 \div 2 = \_\_$
$2 \times 7 = \_\_$	$2 \times ❄ = 14$	$❄ = \_\_$	$14 \div 2 = \_\_$
$2 \times 9 = \_\_$	$2 \times ❄ = 18$	$❄ = \_\_$	$18 \div 2 = \_\_$
$3 \times 5 = \_\_$	$3 \times ❄ = 15$	$❄ = \_\_$	$15 \div 3 = \_\_$
$3 \times 4 = \_\_$	$3 \times ❄ = 12$	$❄ = \_\_$	$12 \div 3 = \_\_$
$3 \times 6 = \_\_$	$3 \times ❄ = 18$	$❄ = \_\_$	$18 \div 3 = \_\_$
$4 \times 5 = \_\_$	$4 \times ❄ = 20$	$❄ = \_\_$	$20 \div 4 = \_\_$
$4 \times 4 = \_\_$	$4 \times ❄ = 16$	$❄ = \_\_$	$16 \div 4 = \_\_$

2. <u>Divide</u> and <u>write</u> the remainder in the parentheses.

$13 \div 5 = \_\_ (\_\_)$ $13 \div 2 = \_\_ (\_\_)$

$13 \div 7 = \_\_ (\_\_)$ $13 \div 3 = \_\_ (\_\_)$ $13 \div 8 = \_\_ (\_\_)$

$13 \div 6 = \_\_ (\_\_)$ $13 \div 9 = \_\_ (\_\_)$ $13 \div 4 = \_\_ (\_\_)$

1. <u>Get</u> Pikachu.

"Outa my way, Pikachu! I'm gonna roll up my sleeves and give you a good HUNT!"

2. <u>Find</u> and <u>circle</u> or <u>cross out</u> the words.

E	I	Z	R	R	Y	C	S	T	W	R	W
A	V	R	J	E	A	F	E	O	O	O	H
Q	E	I	G	P	L	E	O	E	O	J	I
P	H	D	T	A	B	K	D	G	D	W	R
S	I	U	P	C	E	E	A	D	S	P	L
P	R	S	T	A	E	D	N	I	F	I	W
E	D	R	I	B	Y	T	R	P	L	G	I
S	S	P	I	D	G	E	O	T	T	O	N
Y	T	I	L	I	G	A	T	R	Y	V	D
W	I	N	G	S	Y	H	Q	W	P	Z	D
D	D	W	Q	R	A	B	M	X	E	I	V
N	L	O	J	P	H	Z	Q	Q	E	R	R

PIDGEY
CAPTURE
BIRD
TORNADOES
WINGS
WOODS
FLAPS
AGILITY
PIDGEOTTO
PIDGEOT
PROTECTIVE
WHIRLWIND

1. <u>Multiply</u> and <u>divide</u>.

5 × 3 = __	5 × ★ = 15	★ = __	15 ÷ 5 = __
5 × 4 = __	5 × ★ = 20	★ = __	20 ÷ 5 = __
6 × 2 = __	6 × ★ = 12	★ = __	12 ÷ 6 = __
6 × 3 = __	6 × ★ = 18	★ = __	18 ÷ 6 = __
7 × 2 = __	7 × ★ = 14	★ = __	14 ÷ 7 = __
8 × 2 = __	8 × ★ = 16	★ = __	16 ÷ 8 = __
9 × 2 = __	9 × ★ = 18	★ = __	18 ÷ 9 = __
10 × 2 = __	10 × ★ = 20	★ = __	20 ÷ 10 = __

2. <u>Divide</u> and <u>write</u> the remainder in the parentheses.

17 ÷ 2 = __ (__)   17 ÷ 5 = __ (__)

17 ÷ 7 = __ (__)   17 ÷ 8 = __ (__)   17 ÷ 3 = __ (__)

17 ÷ 9 = __ (__)   17 ÷ 6 = __ (__)   17 ÷ 4 = __ (__)

3. Dugtrio triggered 4 earthquakes. Each time it dug 8 miles underground. <u>How many miles</u> did it dig in all?

Answer: _____.

1. <u>Divide</u> and <u>write</u> the ⟦remainder⟧ in the parentheses.

$11 \div 7 =$ __ (__)

$11 \div 3 =$ __ (__)

$11 \div 2 =$ __ (__)

$11 \div 6 =$ __ (__)

$11 \div 5 =$ __ (__)

$11 \div 4 =$ __ (__)

$11 \div 8 =$ __ (__)

$11 \div 9 =$ __ (__)

$15 \div 2 =$ __ (__)

$15 \div 3 =$ __ (__)

$15 \div 8 =$ __ (__)

$15 \div 5 =$ __ (__)

$15 \div 7 =$ __ (__)

$15 \div 4 =$ __ (__)

$15 \div 9 =$ __ (__)

$15 \div 6 =$ __ (__)

2. Meowth tried to capture ⟦27⟧ Pokemon in 27 attacks. It failed each ⟦9th⟧ attack. <u>How many Pokemon</u> did it capture?

Answer: _____.

1. <u>Multiply</u> and <u>divide</u>. <u>Connect</u> the multiplication and division diagrams of the same value.

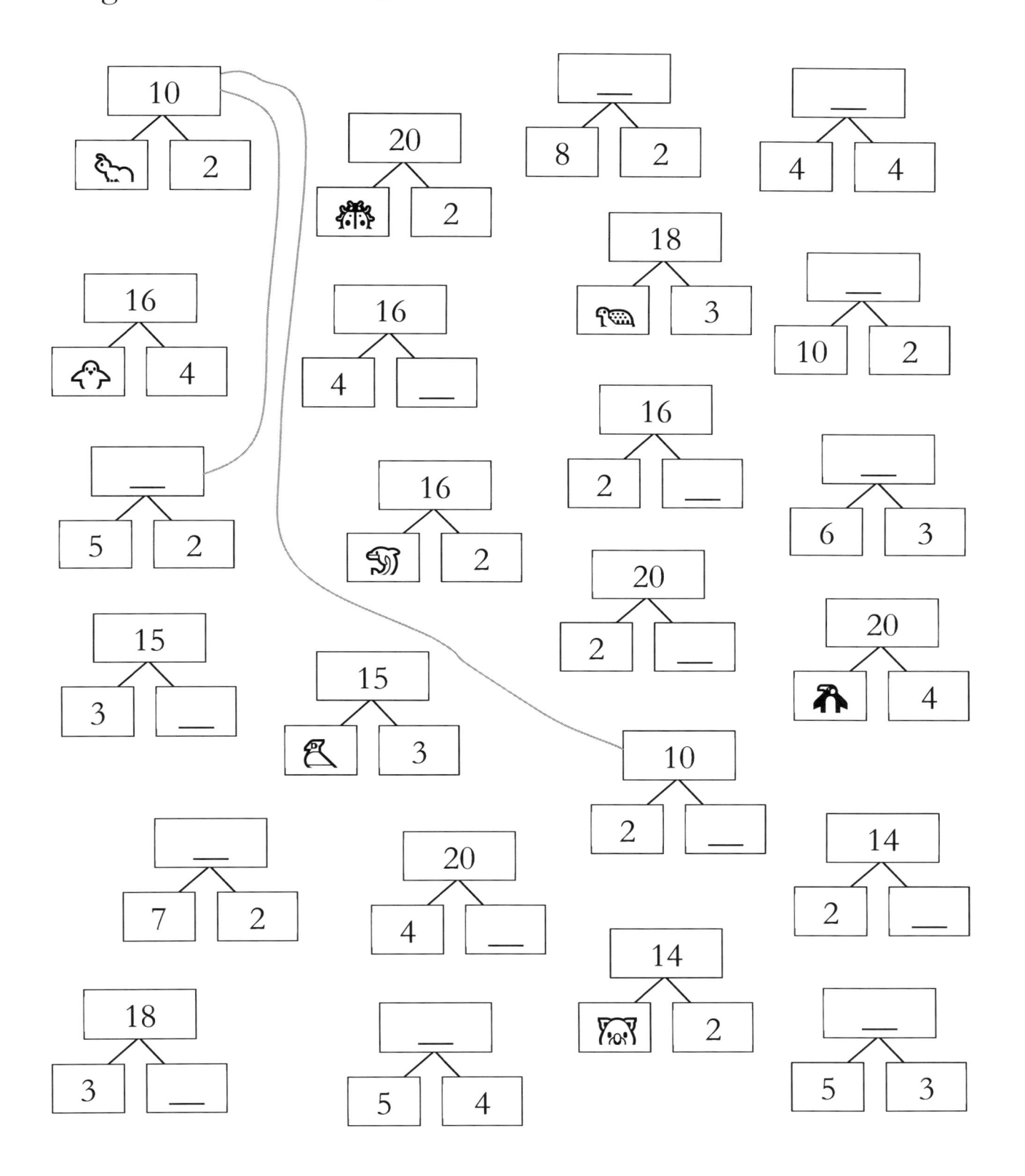

1. <u>Divide</u> and <u>write</u> the remainder in the parentheses.

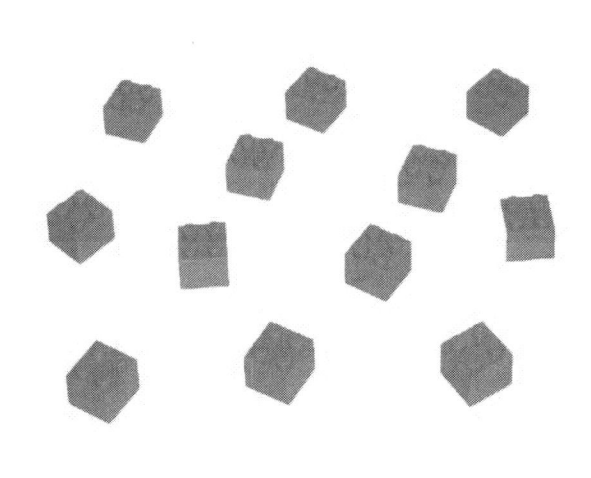

$12 \div 5 =$ __ (__)

$12 \div 2 =$ __ (__)

$12 \div 7 =$ __ (__)

$12 \div 8 =$ __ (__)

$12 \div 9 =$ __ (__)

$12 \div 6 =$ __ (__)

$12 \div 3 =$ __ (__)

$12 \div 4 =$ __ (__)

$18 \div 2 =$ __ (__)

$18 \div 5 =$ __ (__)

$18 \div 3 =$ __ (__)

$18 \div 9 =$ __ (__)

$18 \div 8 =$ __ (__)

$18 \div 4 =$ __ (__)

$18 \div 7 =$ __ (__)

$18 \div 6 =$ __ (__)

2. Seal uses its hard horn to smash through thick ice 3 times a day. <u>How many hits</u> did it use during the week?

Answer: _____.

1. <u>Multiply</u> and <u>divide</u>. <u>Connect</u> the multiplication and division diagrams of the same value.

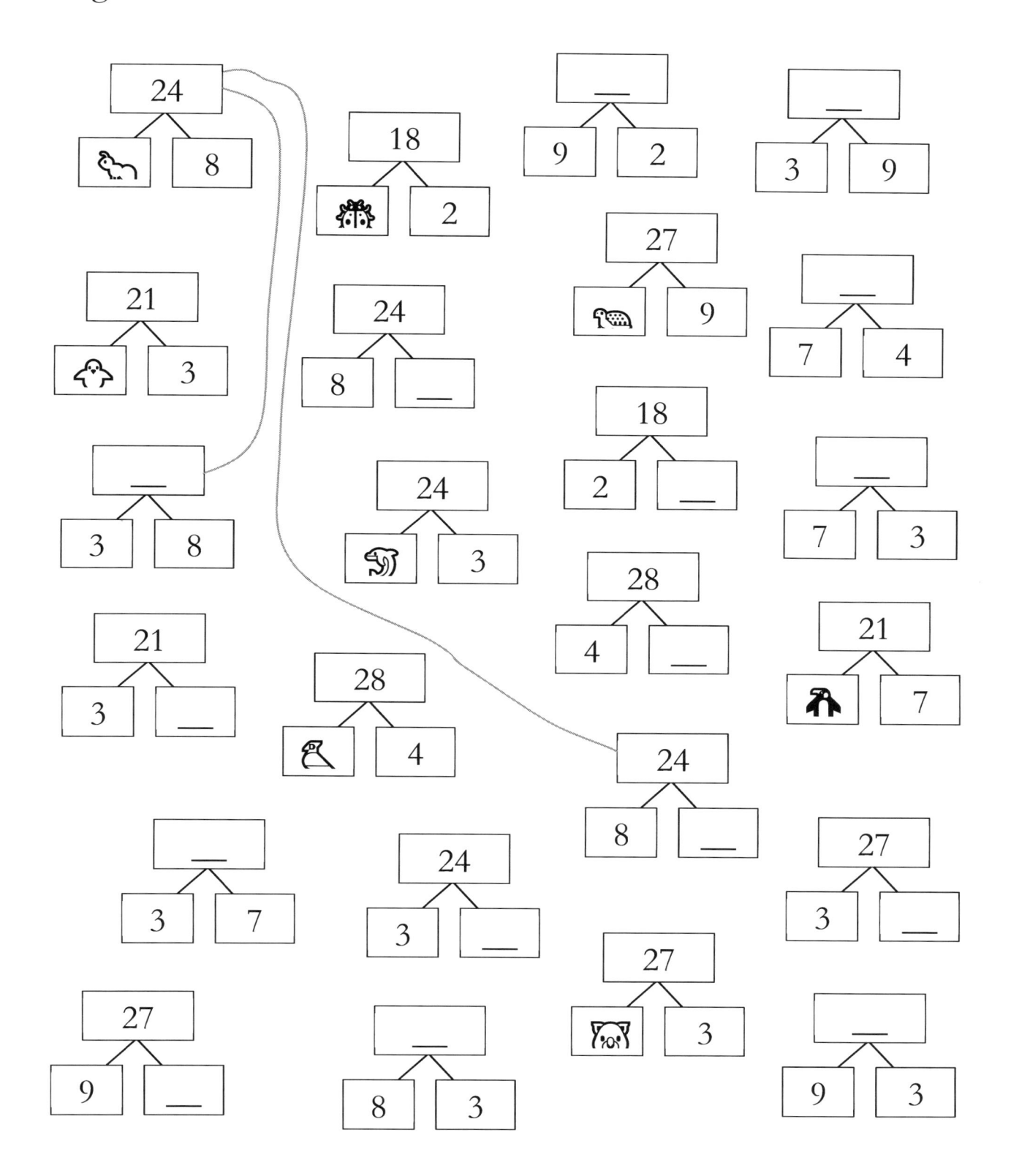

1. <u>Divide</u> and <u>write down</u> the remainder in the parentheses.

$14 \div 2 = \_\_ (\_\_)$

$14 \div 3 = \_\_ (\_\_)$

$14 \div 5 = \_\_ (\_\_)$

$14 \div 9 = \_\_ (\_\_)$

$14 \div 8 = \_\_ (\_\_)$

$14 \div 4 = \_\_ (\_\_)$

$14 \div 6 = \_\_ (\_\_)$

$14 \div 7 = \_\_ (\_\_)$

$19 \div 2 = \_\_ (\_\_)$

$19 \div 3 = \_\_ (\_\_)$

$19 \div 5 = \_\_ (\_\_)$

$19 \div 8 = \_\_ (\_\_)$

$19 \div 9 = \_\_ (\_\_)$

$19 \div 4 = \_\_ (\_\_)$

$19 \div 6 = \_\_ (\_\_)$

$19 \div 7 = \_\_ (\_\_)$

2. Horsea shot 24 blasts of ink in 6 battles. It failed each 2nd battle. <u>How many blasts of ink</u> reached the target?

Answer: _____.

Pokemon Coloring Math Book Multiplication and Division Part 1 Grades 1-4 Ages 6+

1. <u>Get</u> Pikachu.

> Shame on you, Pikachu! Hiding from me! Where d'ya think yer going? Come back!

2. <u>Find</u> and <u>circle</u> or <u>cross out</u> the words.

```
R P A T F W X L S C S S
C X R A W D Y U K I L T
M I X T A I O Z V X U N
D Y E M G N P T F O D I
N F A H O R U S G T G R
M G L S F A I L U R E P
E A I P A E L M S V L T
Z O U R Z P M P E Q H O
P D R A Z A H I H R V O
B R F Y M U K H L M L F
N D S B C K X M F S G G
B P Y D Y S O O B P R E
```

GRIMER
DIRT
MUK
SLIME
SLUDGE
TOXIC
POISONOUS
FOOTPRINTS
DAMAGE
FAILURE
HAZARD

www.stemmindset.com © 2019 STEM mindset, LLC 65

1. <u>Divide</u>.

14 ÷ 7 = __   20 ÷ 4 = __   15 ÷ 3 = __

12 ÷ 3 = __   15 ÷ 5 = __   16 ÷ 2 = __

14 ÷ 2 = __   20 ÷ 5 = __   16 ÷ 8 = __

12 ÷ 6 = __   18 ÷ 3 = __   20 ÷ 2 = __

25 ÷ 5 = __   18 ÷ 6 = __   12 ÷ 2 = __

18 ÷ 2 = __   20 ÷ 10 = __   16 ÷ 16 = __

2. <u>Complete</u> the addition number sentence for each multiplication number sentence and <u>find</u> the value.

2 × 6 = <u>6 + 6 = 12</u>   5 × 3 = _____

4 × 2 = _____   6 × 3 = _____

9 × 3 = _____   7 × 4 = _____

4 × 6 = _____   8 × 3 = _____

7 × 3 = _____   5 × 2 = _____

8 × 4 = _____   6 × 4 = _____

3. Gyarados causes typhoons and sea storms while it gets angry. Last time it caused 3 typhoons and 4 times as many sea storms as typhoons. <u>How many times</u> did it get angry?

Answer: _____.

1. <u>Divide</u> and <u>write</u> the remainder in the parentheses.

$16 \div 2 = $ __ (__)

$16 \div 3 = $ __ (__)

$16 \div 5 = $ __ (__)

$16 \div 8 = $ __ (__)

$16 \div 4 = $ __ (__)

$16 \div 9 = $ __ (__)

$16 \div 7 = $ __ (__)

$16 \div 6 = $ __ (__)

$20 \div 2 = $ __ (__)

$20 \div 5 = $ __ (__)

$20 \div 3 = $ __ (__)

$20 \div 8 = $ __ (__)

$20 \div 9 = $ __ (__)

$20 \div 4 = $ __ (__)

$20 \div 6 = $ __ (__)

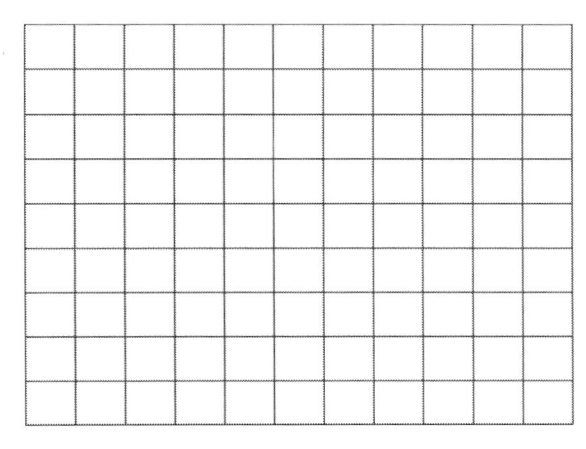

$20 \div 7 = $ __ (__)

2. Draw a 4 × 8 forest where Bulbasaur lives.

Pokemon Coloring Math Book Multiplication and Division Part 1 Grades 1-4 Ages 6+

1. <u>Evaluate</u> each expression.

```
      18            16            14            20            12
  ÷   2   9     ÷   4   __    ÷   2   __    ÷  10   __    ÷   2   __
  -   7   2     -   2   __    +  11   __    +  14   __    -   5   __
  ×   4   8     ×  10   __    ÷   9   __    ÷   4   __    ×   8   __
  ÷   1         ÷   5         ×   6         ÷   2         ÷   4
  ─────────     ─────────     ─────────     ─────────     ─────────
          8            __            __            __            __
```

```
      20            20            15            12             9
  ÷   4   __    ÷   5   __    ÷   5   __    ÷   4   __    ÷   3   __
  -   2   __    -   2   __    +  11   __    +  13   __    -   2   __
  ×   4   __    ×   9   __    ÷   7   __    ÷   4   __    ×  15   __
  ÷   6         ÷   6         ×   6         ×   5         ÷   3
  ─────────     ─────────     ─────────     ─────────     ─────────
         __            __            __            __            __
```

2. <u>How many groups of 3</u> are in $\boxed{15}$ Pikachu? _____

<u>How many groups of 9</u> are in $\boxed{18}$ Pikachu? _____

<u>How many groups of 4</u> are in $\boxed{16}$ Pikachu? _____

<u>How many groups of 2</u> are in 14 Pikachu? _____

© 2019 STEM mindset, LLC www.stemmindset.com

1. <u>Write</u> the missing numbers, <u>multiply</u> and <u>divide</u>.

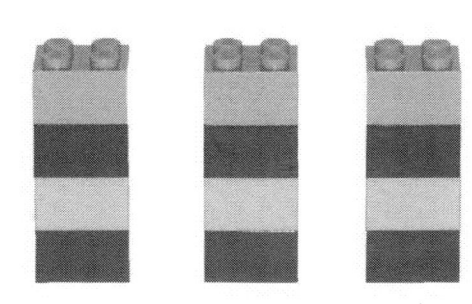

$4 \times 3 =$ __
$12 \div 4 =$ __
$12 \div 3 =$ __

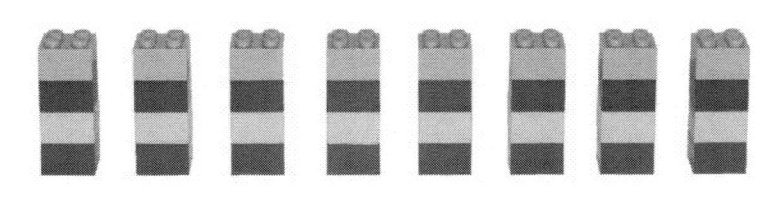

$4 \times 8 =$ __
__ $\div$ __ $=$ __
__ $\div$ __ $=$ __

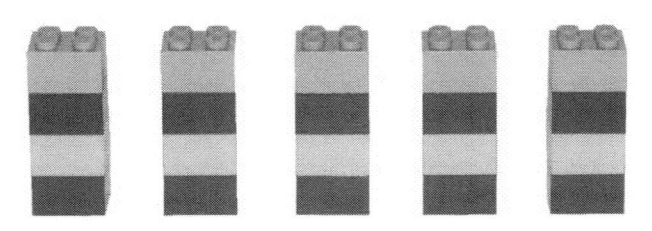

$4 \times 5 =$ __
__ $\div$ __ $=$ __
__ $\div$ __ $=$ __

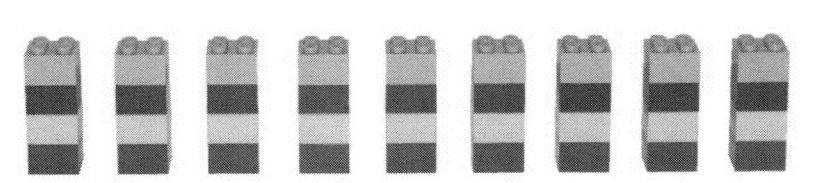

$4 \times 9 =$ __
__ $\div$ __ $=$ __
__ $\div$ __ $=$ __

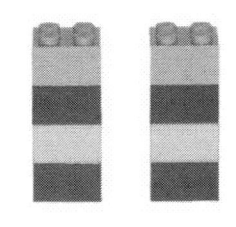

$4 \times 2 =$ __
__ $\div$ __ $=$ __
__ $\div$ __ $=$ __

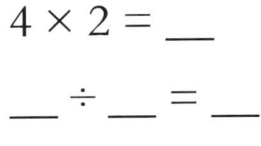

2. I have some numbers and signs: 2, 3, 2, ×, +.

<u>Write</u> the equation that equals one of the answer choices.

A   8   C   12
B 10   D   14

_____

1. <u>Write</u> the missing numbers, <u>multiply</u> and <u>divide</u>.

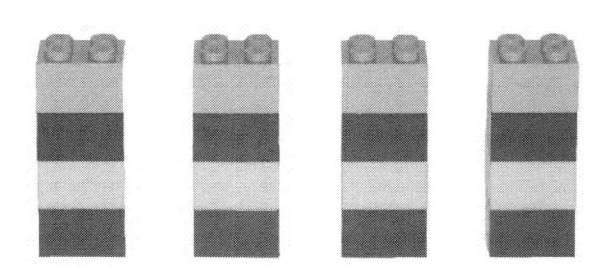

$4 \times 4 =$ __

$16 \div 4 =$ __

$16 \div 4 =$ __

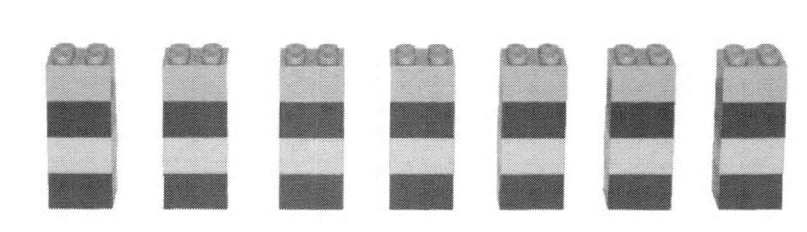

$4 \times 7 =$ __

__ $\div$ __ $=$ __

__ $\div$ __ $=$ __

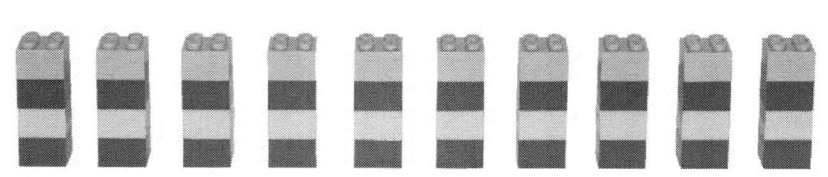

$4 \times 10 =$ __

__ $\div$ __ $=$ __

__ $\div$ __ $=$ __

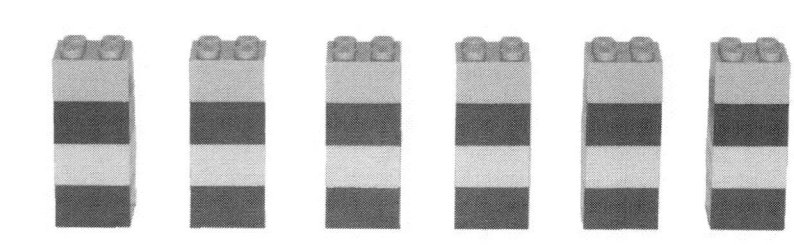

$4 \times 6 =$ __

__ $\div$ __ $=$ __

__ $\div$ __ $=$ __

2. I have some numbers and signs: 7, 3, 5, ×, -.

<u>Write</u> the equation that equals one of the answer choices.

A 5   C 6

B 9   D 21

1. <u>Multiply</u>.

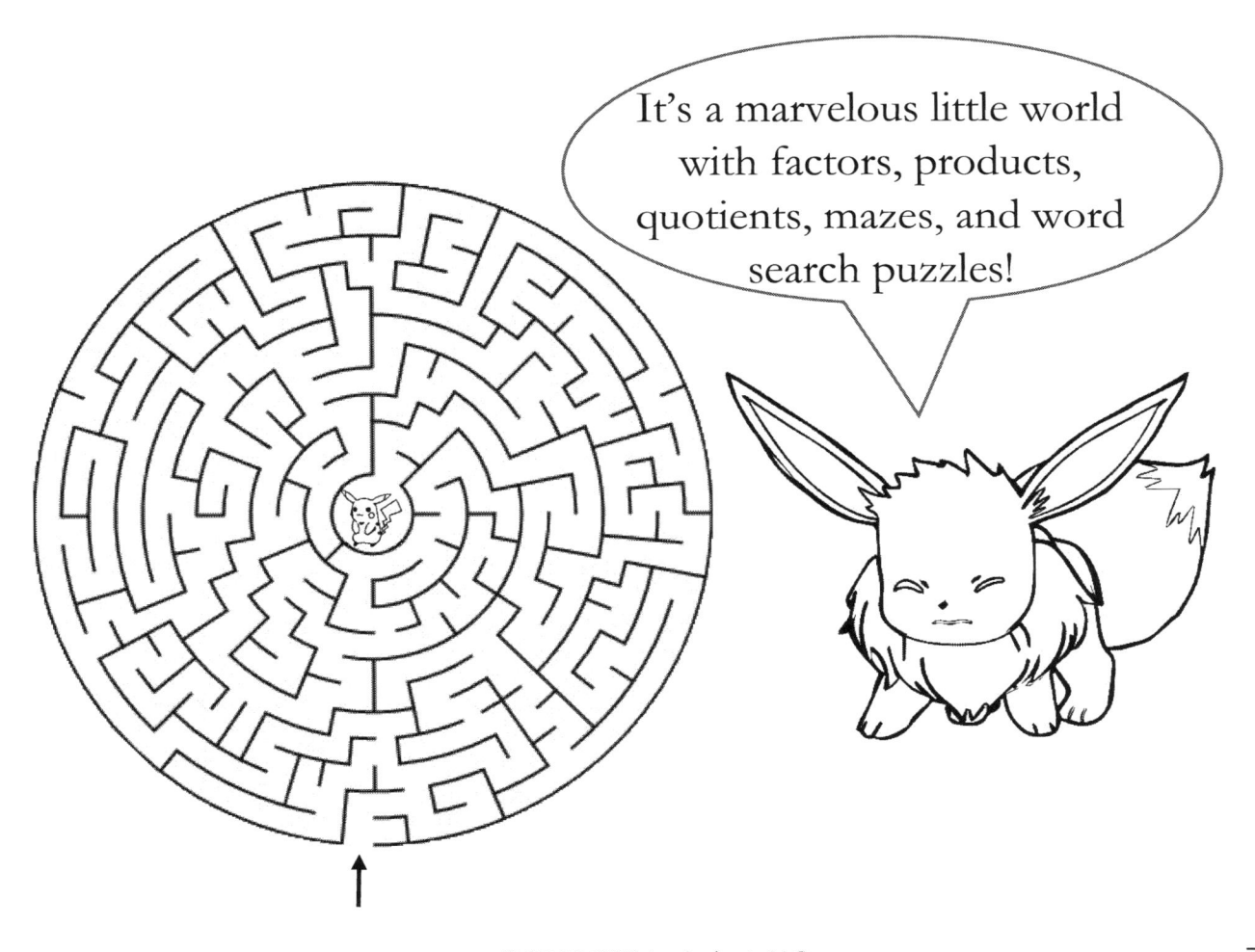

    3   5   8   1   4
× 4   × 4   × 4   × 4   × 4

    6   9   7   2   10
× 4   × 4   × 4   × 4   × 4

It's a marvelous little world with factors, products, quotients, mazes, and word search puzzles!

1. <u>Make</u> 4 rows. X shows the number of columns. <u>Find</u> X. <u>Write</u> the missing numbers, <u>multiply</u> and <u>divide</u>.

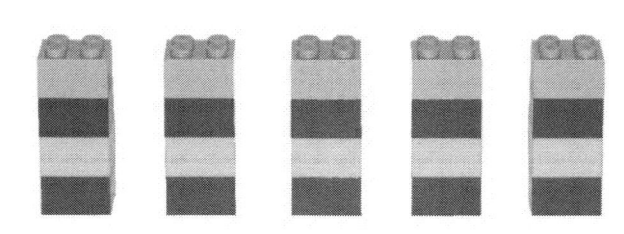

$4 \times X = 20$ X = __

$20 \div 5 =$ __ $20 \div 4 =$ __

__ rows

__ bricks in each row

$4 \times X = 32$ X = __

$32 \div 8 =$ __ $32 \div 4 =$ __

__ rows

__ bricks in each row

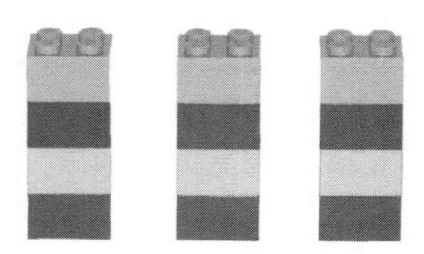

$4 \times X = 12$ X = __

$12 \div 3 =$ __ $12 \div 4 =$ __

__ rows

__ bricks in each row

2. <u>Circle</u> the right answer.

A) $2 \times 5 \times 3$

B) $4 \times 2 \times 3$

a) A is greater than B

b) A is equal to B

c) A is less than B

3. <u>Circle</u> the right answer.

A) $(4 \times 8) - 15$

B) $(5 \times 6) - 14$

a) A is greater than B

b) A is equal to B

c) A is less than B

1. Make 4 rows. X shows the number of columns. Find X. Write the missing numbers, multiply and divide.

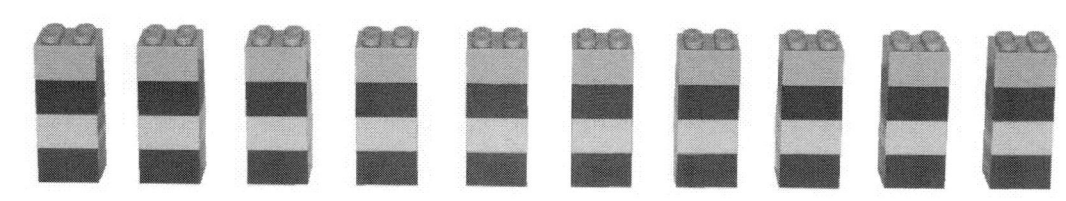

$4 \times X = 40$ $X = \underline{}$

$40 \div 10 = \underline{}$ $40 \div 4 = \underline{}$

__ rows __ bricks in each row

$4 \times X = 16$ $X = \underline{}$

$16 \div 4 = \underline{}$ $16 \div 4 = \underline{}$

__ rows

__ bricks in each row

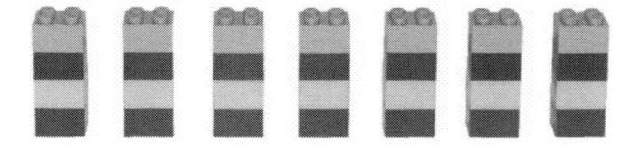

$4 \times X = 28$ $X = \underline{}$

$28 \div 7 = \underline{}$ $28 \div 4 = \underline{}$

__ rows __ bricks in each row

2. Circle the right answer.

A) $3 \times 3 \times 3$

B) $1 \times 6 \times 5$

a) A is greater than B

b) A is equal to B

c) A is less than B

1. Make 4 rows. X shows the number of columns. Find X. Write the missing numbers, multiply and divide.

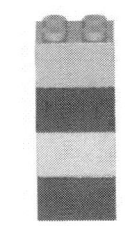

$4 \times X = 24$ $X = \underline{}$

$24 \div 6 = \underline{}$ $24 \div 4 = \underline{}$

__ rows   __ bricks in each row

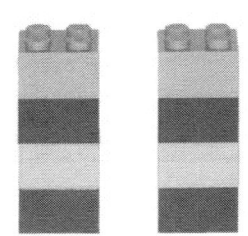

$4 \times X = 8$ $X = \underline{}$

$8 \div 2 = \underline{}$ $8 \div 4 = \underline{}$

__ rows

__ bricks in each row

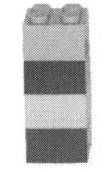

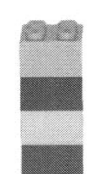

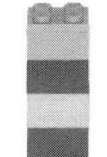

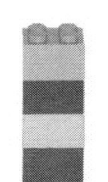

$4 \times X = 36$ $X = \underline{}$

$36 \div 9 = \underline{}$ $36 \div 4 = \underline{}$

__ rows   __ bricks in each row

2. Circle the right answer.

A) $21 \div 7 \times 2$

B) $5 \times 4 \div 2$

a) A is greater than B

b) A is equal to B

c) A is less than B

1. <u>Write</u> the missing numbers.

× __	× __	× __	× __	× __
4	4	4	4	4
4	12	24	40	36

× __	× __	× __	× __	× __
4	4	4	4	4
16	28	8	20	32

"I know my own strength! Pikachu, I'm coming!"

1. <u>Add</u> and <u>multiply</u>. The easiest way to count 4's is counting by groups of three 4's. Three 4's (4 + 4 + 4) equal 12.

<u>4 + 4 + 4</u> = __   3 × 4 = __

<u>4 + 4 + 4</u> + 4 = __   4 × 4 = __

<u>4 + 4 + 4</u> + 4 + 4 = __   5 × 4 = __

<u>4 + 4 + 4</u> + <u>4 + 4 + 4</u> = __   6 × 4 = __

<u>4 + 4 + 4</u> + <u>4 + 4 + 4</u> + 4 = __   7 × 4 = __

<u>4 + 4 + 4</u> + <u>4 + 4 + 4</u> + 4 + 4 = __   8 × 4 = __

<u>4 + 4 + 4</u> + <u>4 + 4 + 4</u> + <u>4 + 4 + 4</u> = __   9 × 4 = __

<u>4 + 4 + 4</u> + <u>4 + 4 + 4</u> + <u>4 + 4 + 4</u> + 4 = __   10 × 4 = __

2. <u>Take</u> __ __ bricks to answer the questions.

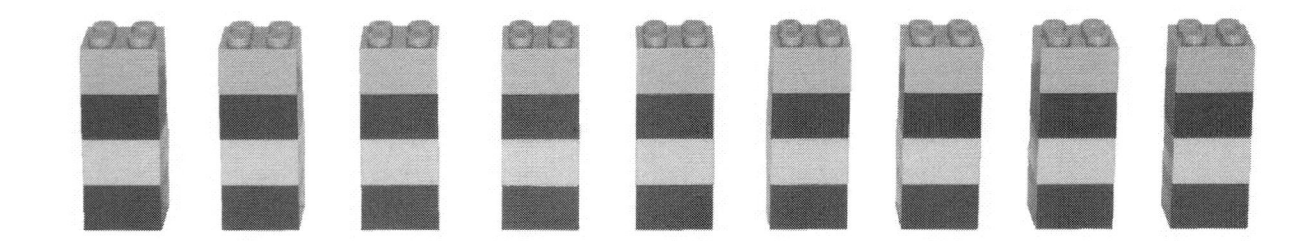

<u>Make</u> 4 rows.

<u>How many</u> columns can you make with 20 bricks?   __

<u>How many</u> columns can you make with the leftover bricks? __

<u>How many</u> columns can you make out of all the bricks?   __.

Pokemon Coloring Math Book Multiplication and Division Part 1 Grades 1-4 Ages 6+

1. <u>Get</u> Pikachu.

Oh, goody, this maze is better than anything! Ho Ho ho ho – I'm glad I got Pikachu!

2. <u>Find</u> and <u>circle</u> or <u>cross out</u> the words.

```
R V R B J S B S T F W S
E I I E X T E R U F V P
D C E X T V E B O L S R
W T E F A A C R O Y A
O R D E M C E B P Q T Y
P E L O I E C Y S E V S
W E E P I N B E L L T I
V B S S A R G B L F B X
A E J U N G L E E B J H
I L D J Q G O Y B T Q M
B E R U T S I O M A I J
Z X V J S S N E U G Y A
```

BELLSPROUT
MOISTURE
ROOTS
WEEPINBELL
SPRAY
POWDER
VICTREEBEL
LEAVES
GRASS
JUNGLE
FLYEATER

www.stemmindset.com © 2019 STEM mindset, LLC 77

1. Write the missing numbers, multiply and divide.

$5 \times 3 =$ __
$15 \div 5 =$ __
$15 \div 3 =$ __

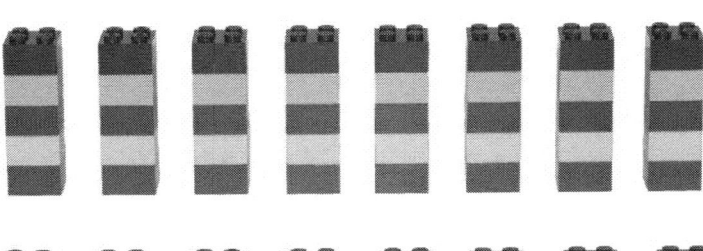

$5 \times 8 =$ __
__ $\div$ __ $=$ __
__ $\div$ __ $=$ __

$5 \times 9 =$ __
__ $\div$ __ $=$ __
__ $\div$ __ $=$ __

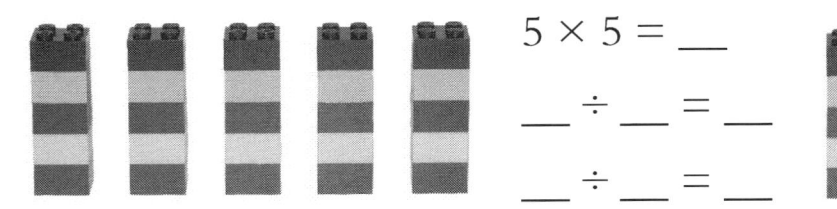

$5 \times 5 =$ __
__ $\div$ __ $=$ __
__ $\div$ __ $=$ __

$5 \times 2 =$ __
__ $\div$ __ $=$ __
__ $\div$ __ $=$ __

2. Circle the right answer.

A) $45 \div 5 \div 3$

B) $36 \div 4 \div 3$

a) A is greater than B

b) A is equal to B

c) A is less than B

1. <u>Write</u> the missing numbers, <u>multiply</u> and <u>divide</u>.

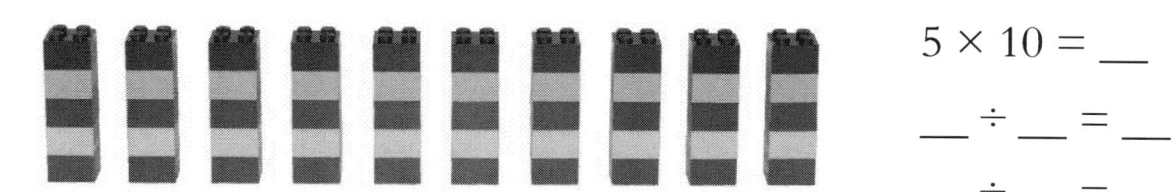

$5 \times 4 = \underline{}$

$20 \div 5 = \underline{}$

$20 \div 4 = \underline{}$

$5 \times 7 = \underline{}$

$\underline{} \div \underline{} = \underline{}$

$\underline{} \div \underline{} = \underline{}$

$5 \times 10 = \underline{}$

$\underline{} \div \underline{} = \underline{}$

$\underline{} \div \underline{} = \underline{}$

$5 \times 6 = \underline{}$

$\underline{} \div \underline{} = \underline{}$

$\underline{} \div \underline{} = \underline{}$

2. <u>Circle</u> the right answer.

   A) $4 \times 2 \times 4$

   B) $2 \times 5 \times 3$

   a) A is greater than B

   b) A is equal to B

   c) A is less than B

1. <u>Multiply</u>.

```
    3          5          8          1          4
×   5      ×   5      ×   5      ×   5      ×   5
_____      _____      _____      _____      _____


    6          9          7          2         10
×   5      ×   5      ×   5      ×   5      ×   5
_____      _____      _____      _____      _____
```

Poor little doomed Pikachu! Come, fella, let's leave now!

1. Make 5 rows. X shows the number of columns. Find X. Write the missing numbers, multiply and divide.

$5 \times X = 20$ X = __

$20 \div 4 =$ __ $20 \div 5 =$ __

__ rows

__ bricks in each row

$5 \times X = 35$ X = __

$35 \div 7 =$ __ $35 \div 5 =$ __

__ rows

__ bricks in each row

$5 \times X = 15$ X = __

$15 \div 3 =$ __ $15 \div 5 =$ __

__ rows

__ bricks in each row

2. Circle the right answer:

I have a series of numbers:
2, 4, 6, …, …, 12, __

What is the next number: __?

a) 15

b) 14

c) 16

1. Make 5 rows. X shows the number of columns. Find X. Write the missing numbers, multiply and divide.

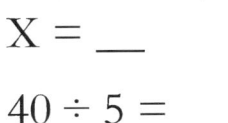

5 × X = 40

40 ÷ 8 = __

__ rows

X = __

40 ÷ 5 = __

__ bricks in each row

5 × X = 10

10 ÷ 2 = __

__ rows

__ bricks in each row

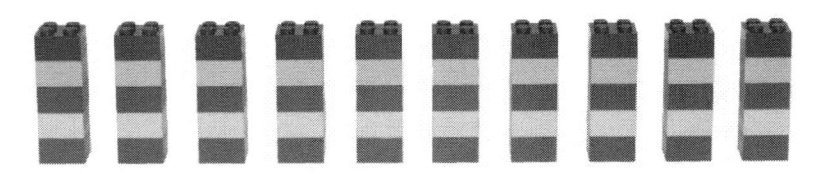

X = __

10 ÷ 5 = __

5 × X = 50

50 ÷ 10 = __

__ rows

X = __

50 ÷ 5 = __

__ bricks in each row

2. Circle the right answer:

I have a series of numbers: 4, 8, 12, (…), (…), (…), ___.

What is the next number: __?

a) 24

b) 32

c) 28

1. Make 5 rows. X shows the number of columns. Find X. Write the missing numbers, multiply and divide.

$5 \times X = 45$ $X = \underline{}$

$45 \div 9 = \underline{}$ $45 \div 5 = \underline{}$

__ rows   __ bricks in each row

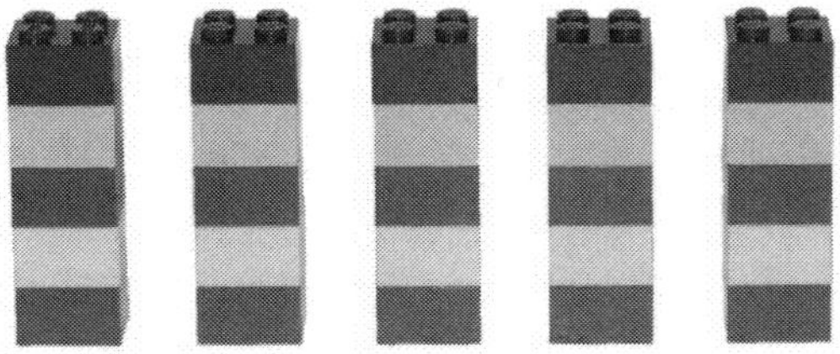

$5 \times X = 25$ $X = \underline{}$

$25 \div 5 = \underline{}$ $25 \div 5 = \underline{}$

__ rows

__ bricks in each row

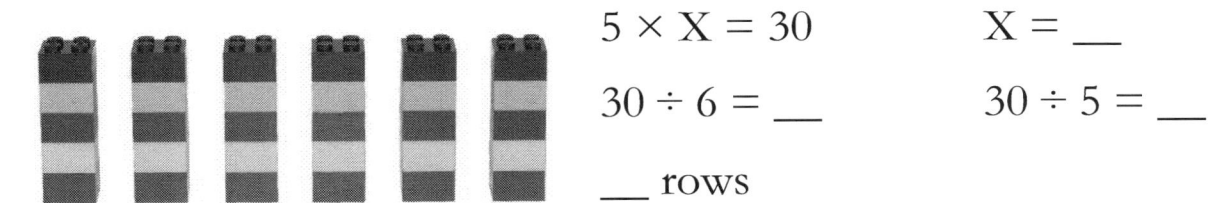

$5 \times X = 30$ $X = \underline{}$

$30 \div 6 = \underline{}$ $30 \div 5 = \underline{}$

__ rows

__ bricks in each row

2. I have some numbers and signs: 5, 8, 4, ×, ÷.

Make an equation that equals one the answer choices. Circle the right answer.

A   3   C   8

B   10   D   6

1. <u>Write</u> the missing numbers.

$$\begin{array}{r} \_\_ \\ \times\ 5 \\ \hline 5 \end{array} \qquad \begin{array}{r} \_\_ \\ \times\ 5 \\ \hline 20 \end{array} \qquad \begin{array}{r} \_\_ \\ \times\ 5 \\ \hline 35 \end{array} \qquad \begin{array}{r} \_\_ \\ \times\ 5 \\ \hline 15 \end{array} \qquad \begin{array}{r} \_\_ \\ \times\ 5 \\ \hline 25 \end{array}$$

$$\begin{array}{r} \_\_ \\ \times\ 5 \\ \hline 40 \end{array} \qquad \begin{array}{r} \_\_ \\ \times\ 5 \\ \hline 50 \end{array} \qquad \begin{array}{r} \_\_ \\ \times\ 5 \\ \hline 10 \end{array} \qquad \begin{array}{r} \_\_ \\ \times\ 5 \\ \hline 45 \end{array} \qquad \begin{array}{r} \_\_ \\ \times\ 5 \\ \hline 30 \end{array}$$

Yoo Hoo, Pikachu! Do you mind telling us how you're solving all these mazes?

Pokemon Coloring Math Book Multiplication and Division Part 1 Grades 1-4 Ages 6+

1. <u>Find</u> the value. The easiest way to count 5's is counting by groups of two 5's. Two 5's (5 + 5) equal 10.

<u>5 + 5</u> = __   2 × 5 = __

<u>5 + 5</u> + 5 = __   3 × 5 = __

<u>5 + 5</u> + <u>5 + 5</u> = __   4 × 5 = __

<u>5 + 5</u> + <u>5 + 5</u> + 5 = __   5 × 5 = __

<u>5 + 5</u> + <u>5 + 5</u> + <u>5 + 5</u> = __   6 × 5 = __

<u>5 + 5</u> + <u>5 + 5</u> + <u>5 + 5</u> + 5 = __   7 × 5 = __

<u>5 + 5</u> + <u>5 + 5</u> + <u>5 + 5</u> + <u>5 + 5</u> = __   8 × 5 = __

<u>5 + 5</u> + <u>5 + 5</u> + <u>5 + 5</u> + <u>5 + 5</u> + 5 = __ 9 × 5 = __

<u>5 + 5</u> + <u>5 + 5</u> + <u>5 + 5</u> + <u>5 + 5</u> + <u>5 + 5</u> = __   10 × 5 = __

2. <u>Take</u> __ __ bricks to answer the questions.

<u>Make</u> 5 rows.

<u>How many</u> columns can you make with 25 bricks?   __

<u>How many</u> columns can you make with the leftover bricks?   __

<u>How many</u> columns can you make out of all the bricks?   __.

Pokemon Coloring Math Book Multiplication and Division Part 1 Grades 1-4 Ages 6+

My-My-Who'd believe such a little monster capable of hiding so well!

1. <u>Find</u> and <u>circle</u> or <u>cross out</u> the words.

E	N	B	M	D	Y	S	S	M	V	Y	S
N	L	M	U	S	A	T	K	N	U	L	B
A	J	D	M	T	I	P	O	P	I	H	E
V	T	W	E	N	T	O	F	A	X	E	V
T	X	T	G	E	C	E	T	D	J	A	O
S	Y	E	A	O	W	D	R	N	D	D	L
E	R	A	C	C	N	E	O	F	B	H	U
R	L	L	N	A	K	T	P	V	L	J	T
O	S	W	S	K	A	K	U	N	A	Y	I
F	U	G	D	E	H	C	T	A	H	B	O
Z	E	C	E	B	X	P	A	X	O	P	N
L	C	A	T	E	R	P	I	L	L	A	R

WEEDLE
STINGER
FOREST
HEAD
CATERPILLAR
BUTTERFLY
COCOON
EVOLUTION
KAKUNA
HATCHED
ATTACK
LEGS AND TAILS

1. <u>Multiply</u>.

$2 \times 2 =$ __ $2 \times 4 =$ __ $2 \times 6 =$ __

$2 \times 8 =$ __ $2 \times 10 =$ __

$5 \times 2 =$ __ $5 \times 4 =$ __ $5 \times 6 =$ __

$5 \times 8 =$ __ $5 \times 10 =$ __

$3 \times 2 =$ __ $3 \times 4 =$ __ $3 \times 6 =$ __

$3 \times 8 =$ __ $3 \times 10 =$ __

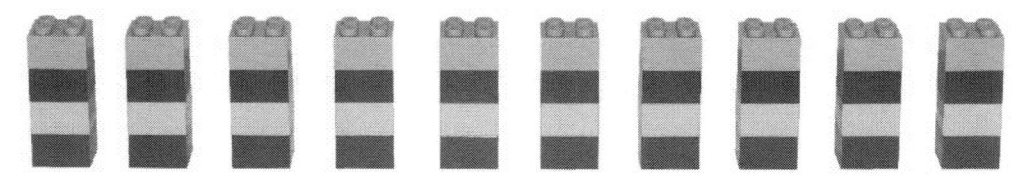

$4 \times 2 =$ __ $4 \times 4 =$ __ $4 \times 6 =$ __

$4 \times 8 =$ __ $4 \times 10 =$ __

2. I have some numbers and signs: 6, 6, 9, ×, ÷.

<u>Make</u> an equation that equals one the answer choices. <u>Circle</u> the right answer.

A 2   B 5   C 4   D 10

1. <u>Multiply</u>.

$2 \times 1 =$ __ $2 \times 3 =$ __ $2 \times 5 =$ __

$2 \times 7 =$ __ $2 \times 9 =$ __

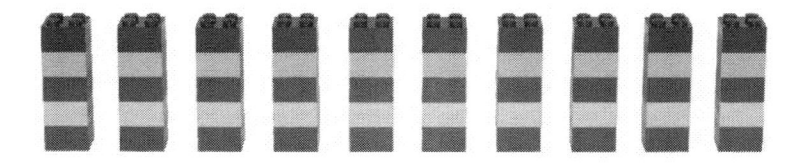

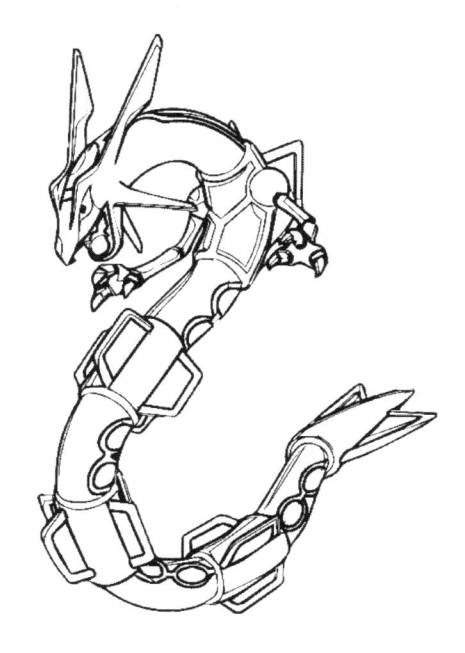

$5 \times 1 =$ __

$5 \times 3 =$ __ $5 \times 5 =$ __

$5 \times 7 =$ __ $5 \times 9 =$ __

$3 \times 1 =$ __ $3 \times 3 =$ __ $3 \times 5 =$ __

$3 \times 7 =$ __ $3 \times 9 =$ __

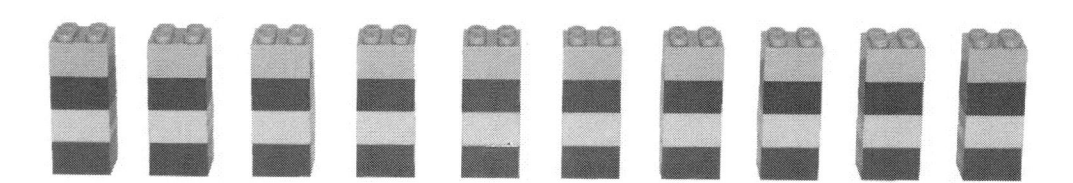

$4 \times 1 =$ __ $4 \times 3 =$ __ $4 \times 5 =$ __

$4 \times 7 =$ __ $4 \times 9 =$ __

1. <u>Explain</u>: 12 is divided by 2. <u>Use</u> bricks, cubes, or sticks to answer the question.

It means _____

_____.

2. <u>Explain</u>: 24 is divided by 4. <u>Use</u> bricks, cubes, or sticks to answer the question.

It means _____

_____.

3. <u>Explain</u>: 36 is divided by 9. <u>Use</u> bricks, cubes, or sticks to answer the question.

It means _____

_____.

4. <u>Explain</u>: 35 is divided by 5. <u>Use</u> bricks, cubes, or sticks to answer the question.

It means _____

_____.

2. I have some numbers and signs: 2, 9, 3, ×, ÷.

<u>Make</u> an equation that equals one the answer choices. <u>Circle</u> the right answer.

A   7   C   9

B   5   D   6

1. <u>Evaluate</u> each expression. First, <u>start</u> with the parentheses (brackets), then, addition or subtraction from left to right! Or first, <u>start</u> with the parentheses (brackets), then, multiplication or subtraction. <u>Indicate</u> order of operations.

$(24 \overset{1}{\div} 3) \overset{4}{+} (15 \overset{2}{\div} 3) \overset{5}{-} (32 \overset{3}{\div} 4) = \underline{}$ $\qquad$ $(18 \overset{1}{\div} 3) \overset{3}{\div} (12 \overset{2}{\div} 4) = \underline{}$

$(20 \div 4) + (21 \div 3) - (12 \div 4) = \underline{}$ $\qquad$ $(27 \div 3) \div (9 \div 3) = \underline{}$

$(12 \div 3) + (36 \div 4) - (18 \div 3) = \underline{}$ $\qquad$ $(28 \div 4) \times (8 \div 4) = \underline{}$

$(27 \div 3) + (16 \div 4) - (9 \div 3) = \underline{}$ $\qquad$ $(20 \div 4) \times (16 \div 4) = \underline{}$

$(28 \div 4) + (18 \div 3) - (8 \div 4) = \underline{}$ $\qquad$ $(36 \div 4) \times (6 \div 3) = \underline{}$

$(21 \div 3) + (12 \div 4) - (24 \div 4) = \underline{}$ $\qquad$ $(32 \div 4) \div (12 \div 3) = \underline{}$

$(35 \div 5) + (15 \div 5) - (32 \div 4) = \underline{}$ $\qquad$ $(45 \div 5) \div (12 \div 4) = \underline{}$

$(25 \div 5) + (20 \div 4) - (10 \div 5) = \underline{}$ $\qquad$ $(30 \div 5) \div (8 \div 4) = \underline{}$

$(30 \div 5) + (28 \div 4) - (45 \div 5) = \underline{}$ $\qquad$ $(40 \div 5) \times (16 \div 4) = \underline{}$

$(35 \div 5) + (15 \div 5) - (18 \div 3) = \underline{}$ $\qquad$ $(45 \div 5) \div (27 \div 3) = \underline{}$

$(25 \div 5) + (21 \div 3) - (10 \div 5) = \underline{}$ $\qquad$ $(30 \div 5) \div (9 \div 3) = \underline{}$

$(30 \div 5) + (18 \div 3) - (45 \div 5) = \underline{}$ $\qquad$ $(40 \div 5) \times (12 \div 3) = \underline{}$

1. Write the missing numbers.

Write all even numbers from 1 up to 20.

___, ___, ___, ___, ___, ___, ___, ___, ___, ___.

Write all odd numbers from 1 up to 20.

___, ___, ___, ___, ___, ___, ___, ___, ___, ___.

Write all the numbers from 1 up to 20, divisible by 2.

___, ___, ___, ___, ___, ___, ___, ___, ___, ___.

Write all the numbers from 1 up to 30, divisible by 3.

___, ___, ___, ___, ___, ___, ___, ___, ___, ___.

Write all the numbers from 1 up to 40, divisible by 4.

___, ___, ___, ___, ___, ___, ___, ___, ___, ___.

Write all the numbers from 1 up to 50, divisible by 5.

___, ___, ___, ___, ___, ___, ___, ___, ___, ___.

Pokemon Coloring Math Book Multiplication and Division Part 1 Grades 1-4 Ages 6+

1. <u>Get</u> Pikachu.

"Now, half the maze is yours, and half is mine – you take all the paths, I take Pikachu!"

2. <u>Find</u> and <u>circle</u> or <u>cross out</u> the words.

```
P G E H H R R I U F E D
S A E E O N O L L R R L
N A R B L T J A I A H W
O E F A T T M F Z D U O
O M R Q L E T I H B R R
Z J E L S Y R A O M M G
I S T J P A Z V B Z Z P
N J T V H E J E D U S T
G J U C B O U L D E R S
G A B R I C N O I T O P
J X U K O H Z B E C W U
G I H L P K S J Q K U B
```

CHARIZARD
HOT
FIRE
GROWL
FLAMES
BOULDERS
BUTTERFREE
DUST
POTION
PARALYZE
BATTLE
SNOOZING

© 2019 STEM mindset, LLC   www.stemmindset.com

1. <u>Multiply</u> and <u>divide</u>. <u>Connect</u> the multiplication and division diagrams of the same value.

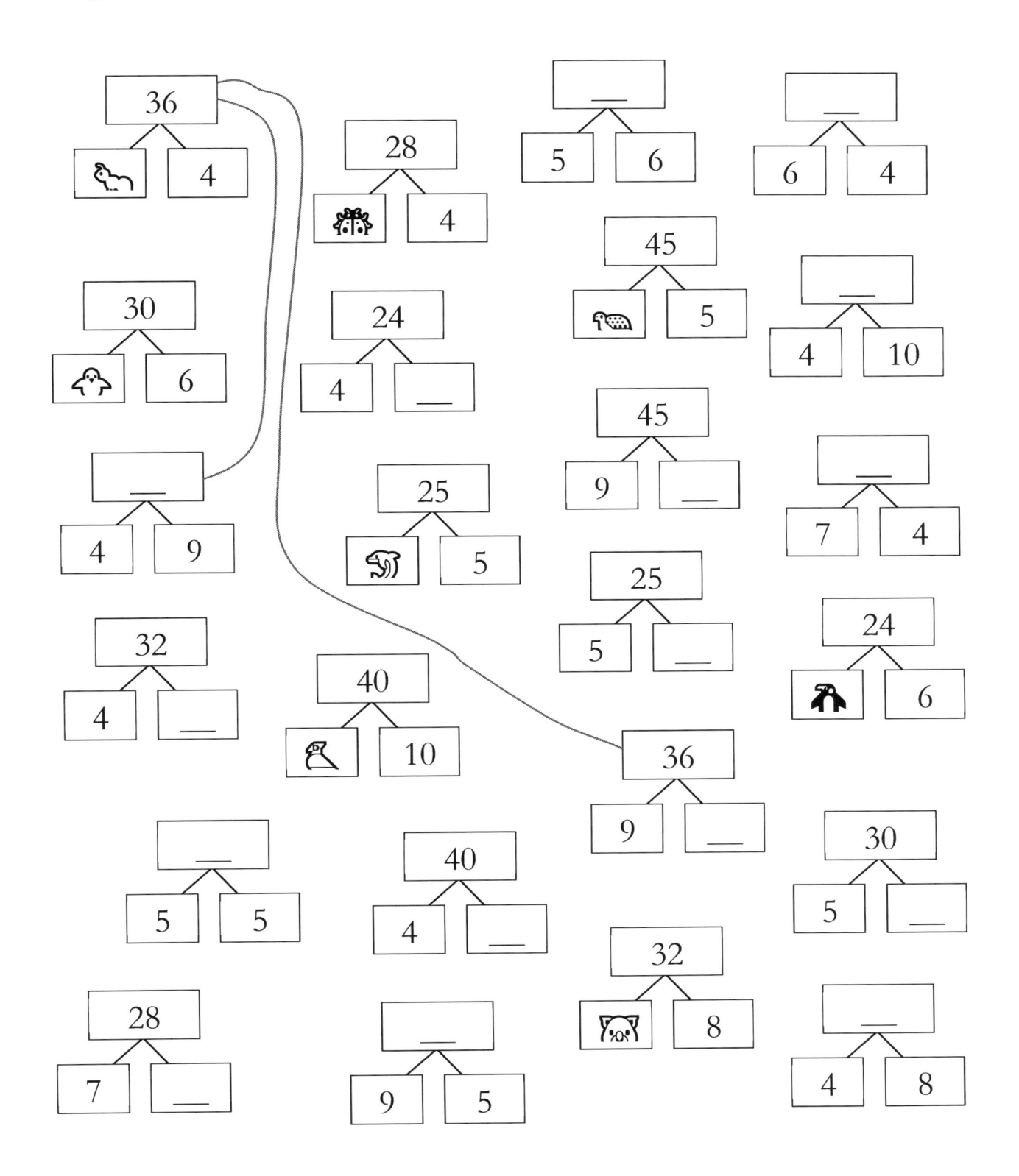

1. <u>Complete</u> the addition number sentence for each multiplication number sentence and <u>evaluate</u> each expression.

3 × 4 = <u>3 + 3 + 3 + 3 = 12</u>   8 × 4 = _____

10 × 2 = _____   4 × 7 = _____

4 × 4 = _____   3 × 2 = _____

2 × 9 = _____   5 × 4 = _____

2. <u>Evaluate</u> each expression. <u>Indicate</u> order of operations.

	1 2	1 2
4 × 5 = __	(4 × 3) + 23 = __	(4 × 5) − 8 = __
4 × 2 = __	(4 × 4) + 19 = __	(4 × 10) − 24 = __
4 × 10 = __	(4 × 7) + 6 = __	(4 × 9) − 18 = __
4 × 6 = __	(4 × 9) + 7 = __	(4 × 8) − 23 = __
5 × 5 = __	(5 × 3) + 15 = __	(5 × 6) − 24 = __
5 × 2 = __	(5 × 4) + 7 = __	(5 × 10) − 33 = __
5 × 10 = __	(5 × 6) + 24 = __	(5 × 9) − 17 = __
5 × 5 = __	(5 × 8) + 32 = __	(5 × 7) − 28 = __

1. <u>Multiply</u>.

5 × 4 = __   5 × 8 = __   5 × 2 = __

5 × 5 = __   5 × 10 = __   5 × 1 = __

2 × 4 = __   2 × 8 = __   2 × 2 = __

2 × 5 = __   2 × 10 = __   2 × 1 = __

4 × 4 = __   4 × 8 = __   4 × 2 = __

4 × 5 = __   4 × 10 = __   4 × 1 = __

3 × 4 = __   3 × 8 = __   3 × 2 = __

3 × 5 = __   3 × 10 = __   3 × 1 = __

5 × 3 = __   5 × 7 = __   5 × 9 = __

2 × 3 = __   2 × 7 = __   2 × 9 = __

4 × 3 = __   4 × 7 = __   4 × 9 = __

3 × 3 = __   3 × 7 = __   3 × 9 = __

1. <u>Divide</u>.

25 ÷ 5 = ___   50 ÷ 5 = ___   10 ÷ 5 = ___

20 ÷ 5 = ___   30 ÷ 5 = ___   40 ÷ 5 = ___

15 ÷ 5 = ___   35 ÷ 5 = ___   45 ÷ 5 = ___

10 ÷ 2 = ___   20 ÷ 2 = ___   4 ÷ 2 = ___

8 ÷ 2 = ___   12 ÷ 2 = ___   16 ÷ 2 = ___

6 ÷ 2 = ___   14 ÷ 2 = ___   18 ÷ 2 = ___

20 ÷ 4 = ___   40 ÷ 4 = ___   8 ÷ 4 = ___

16 ÷ 4 = ___   24 ÷ 4 = ___   32 ÷ 4 = ___

12 ÷ 4 = ___   28 ÷ 4 = ___   36 ÷ 4 = ___

15 ÷ 3 = ___   30 ÷ 3 = ___   6 ÷ 3 = ___

12 ÷ 3 = ___   18 ÷ 3 = ___   24 ÷ 3 = ___

9 ÷ 3 = ___   21 ÷ 3 = ___   27 ÷ 3 = ___

Pokemon Coloring Math Book Multiplication and Division Part 1 Grades 1-4 Ages 6+

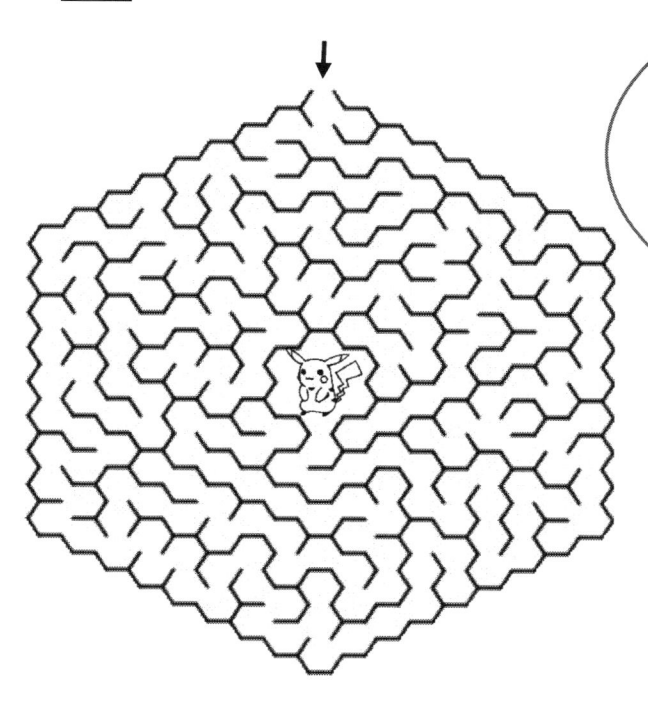

1. <u>Get</u> Pikachu.

> Yeh…I guess you're RIGHT about math, Pikachu! There's always plenty of fun in multiplying and dividing!

2. <u>Find</u> and <u>circle</u> or <u>cross out</u> the words.

```
T Q C E Z Y G G X S R
R Q A V N N A H B U E
A E Q V I R S L S R N
M B G Z C Y P E E P I
S C A H V Z I N D R A
W M O U N T A I N I R
A M J U I D S R T S T
P X Q L S R E D A E L
L A I E L B A U L A V
R B P O P U L A R U W
A F R I D J X X P M I
```

SMART

TRAINER

AMAZING

ABILITIES

POPULAR

MOUNTAIN

SURPRISE

GARCHOMP

VALUABLE

LEADERS

www.stemmindset.com © 2019 STEM mindset, LLC

Page 7

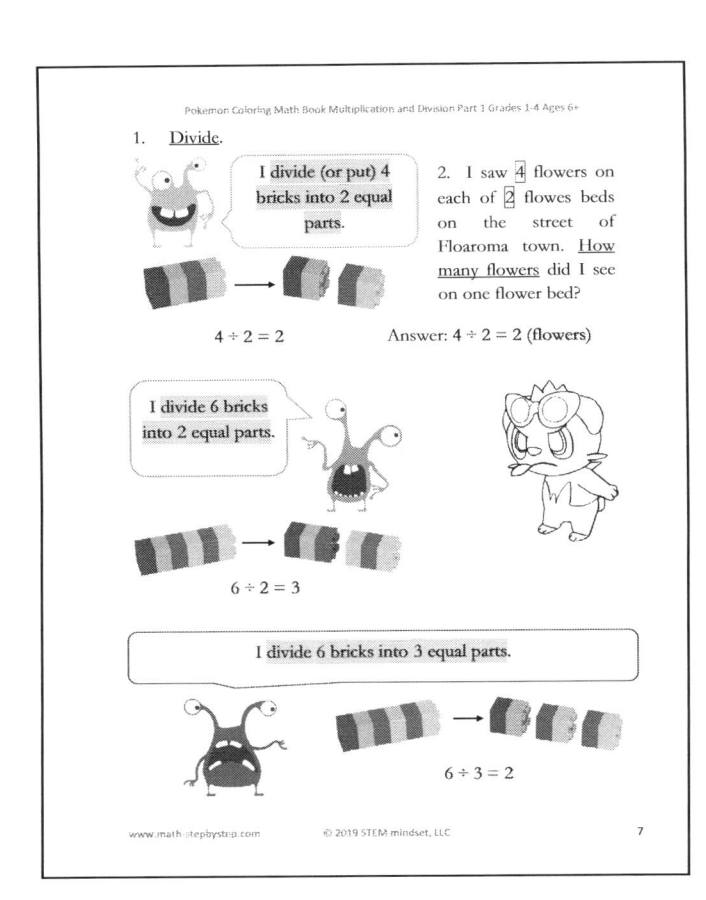

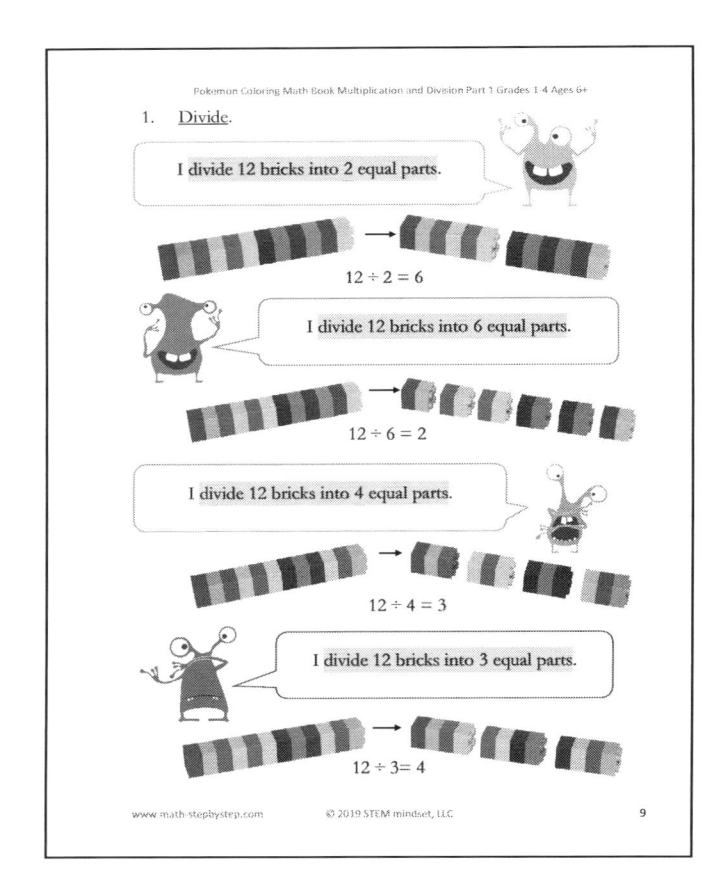

1. Divide.

"I divide (or put) 4 bricks into 2 equal parts."

2. I saw 4 flowers on each of 2 flowes beds on the street of Floaroma town. How many flowers did I see on one flower bed?

$4 \div 2 = 2$

Answer: $4 \div 2 = 2$ (flowers)

"I divide 6 bricks into 2 equal parts."

$6 \div 2 = 3$

"I divide 6 bricks into 3 equal parts."

$6 \div 3 = 2$

Page 8

1. Divide.

"I divide 8 bricks into 2 equal parts."

$8 \div 2 = 4$

"I divide 8 bricks into 4 equal parts."

$8 \div 4 = 2$

"I divide 10 bricks into 2 equal parts."

$10 \div 2 = 5$

"I divide 10 bricks into 5 equal parts."

$10 \div 5 = 2$

Page 9

1. Divide.

"I divide 12 bricks into 2 equal parts."

$12 \div 2 = 6$

"I divide 12 bricks into 6 equal parts."

$12 \div 6 = 2$

"I divide 12 bricks into 4 equal parts."

$12 \div 4 = 3$

"I divide 12 bricks into 3 equal parts."

$12 \div 3 = 4$

Page 10

1. Divide.

"I divide 14 bricks into 2 equal parts."

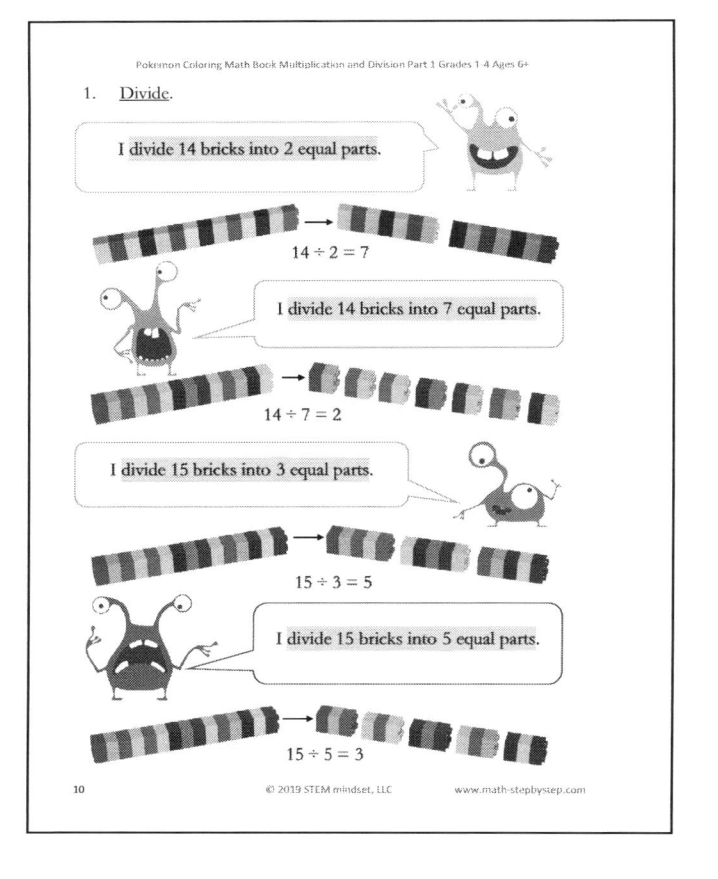

$14 \div 2 = 7$

"I divide 14 bricks into 7 equal parts."

$14 \div 7 = 2$

"I divide 15 bricks into 3 equal parts."

$15 \div 3 = 5$

"I divide 15 bricks into 5 equal parts."

$15 \div 5 = 3$

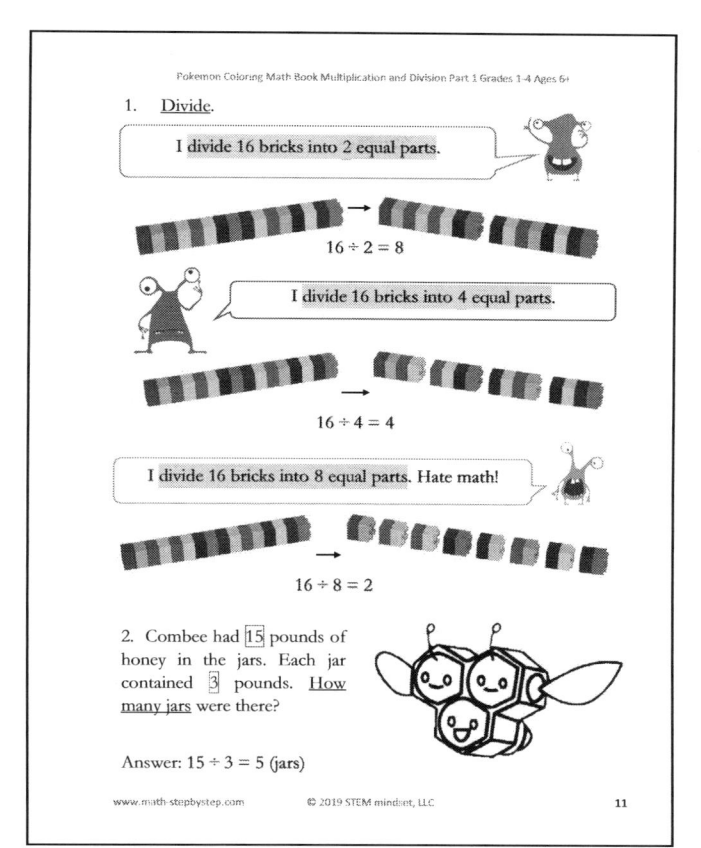

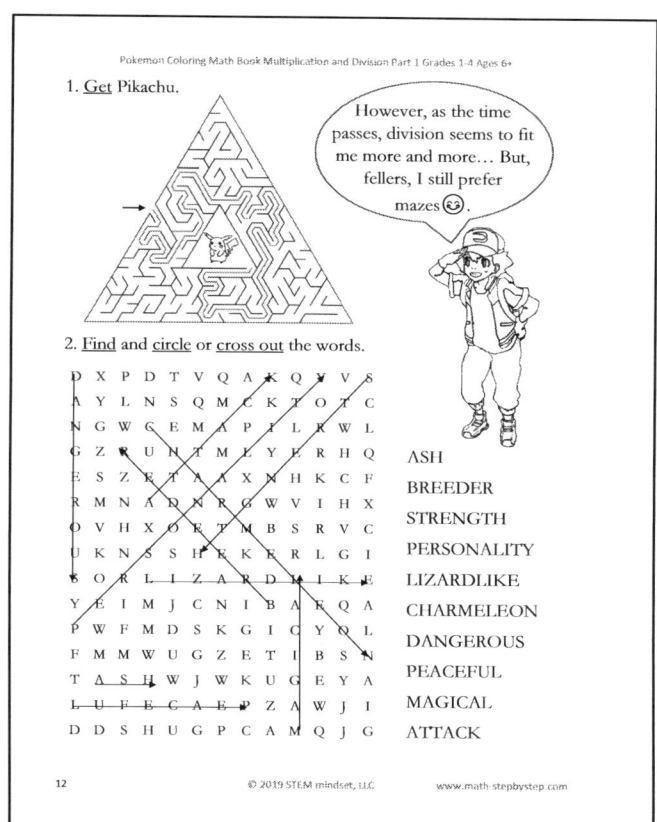

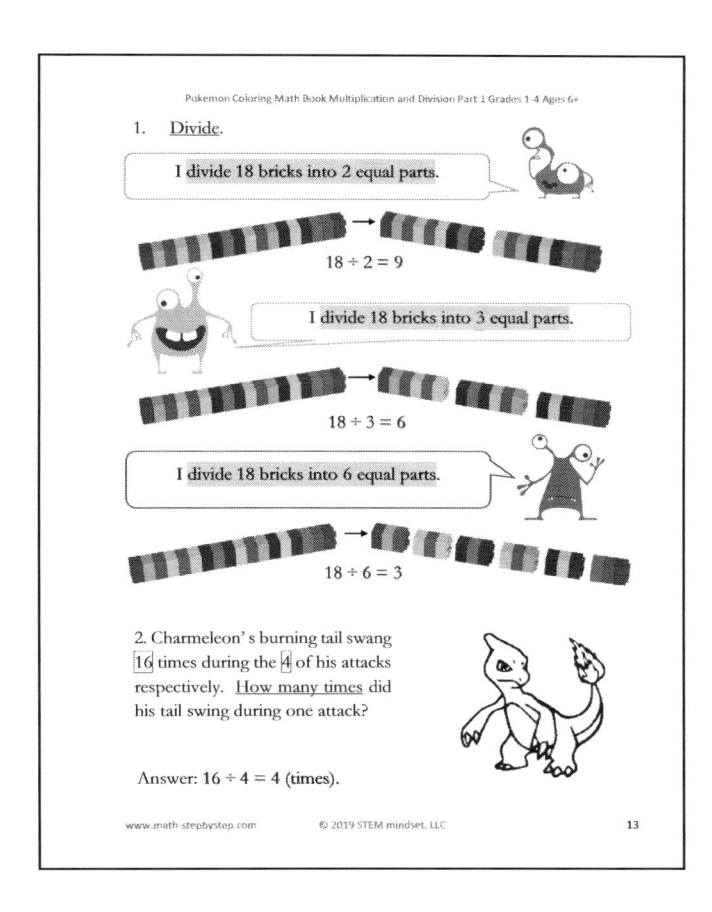

Pokemon Coloring Math Book Multiplication and Division Part 1 Grades 1-4 Ages 6+

1. <u>Divide</u>.

   I divide 16 bricks into 2 equal parts.

   $16 \div 2 = 8$

   I divide 16 bricks into 4 equal parts.

   $16 \div 4 = 4$

   I divide 16 bricks into 8 equal parts. Hate math!

   $16 \div 8 = 2$

2. Combee had 15 pounds of honey in the jars. Each jar contained 3 pounds. <u>How many jars</u> were there?

   Answer: $15 \div 3 = 5$ (jars)

Pokemon Coloring Math Book Multiplication and Division Part 1 Grades 1-4 Ages 6+

1. <u>Get</u> Pikachu.

   However, as the time passes, division seems to fit me more and more... But, fellers, I still prefer mazes 😊.

2. <u>Find</u> and <u>circle</u> or <u>cross out</u> the words.

   ASH
   BREEDER
   STRENGTH
   PERSONALITY
   LIZARDLIKE
   CHARMELEON
   DANGEROUS
   PEACEFUL
   MAGICAL
   ATTACK

Pokemon Coloring Math Book Multiplication and Division Part 1 Grades 1-4 Ages 6+

1. <u>Divide</u>.

   I divide 18 bricks into 2 equal parts.

   $18 \div 2 = 9$

   I divide 18 bricks into 3 equal parts.

   $18 \div 3 = 6$

   I divide 18 bricks into 6 equal parts.

   $18 \div 6 = 3$

2. Charmeleon's burning tail swang 16 times during the 4 of his attacks respectively. <u>How many times</u> did his tail swing during one attack?

   Answer: $16 \div 4 = 4$ (times).

Pokemon Coloring Math Book Multiplication and Division Part 1 Grades 1-4 Ages 6+

1. <u>Divide</u>.

   I divide 20 bricks into 4 equal parts.

   $20 \div 4 = 5$

   I divide 20 bricks into 5 equal parts.

   $20 \div 5 = 4$

   I divide 20 bricks into 2 equal parts.

   $20 \div 2 = 10$

   I divide 20 bricks into 10 equal parts.

   $20 \div 10 = 2$

Page 15

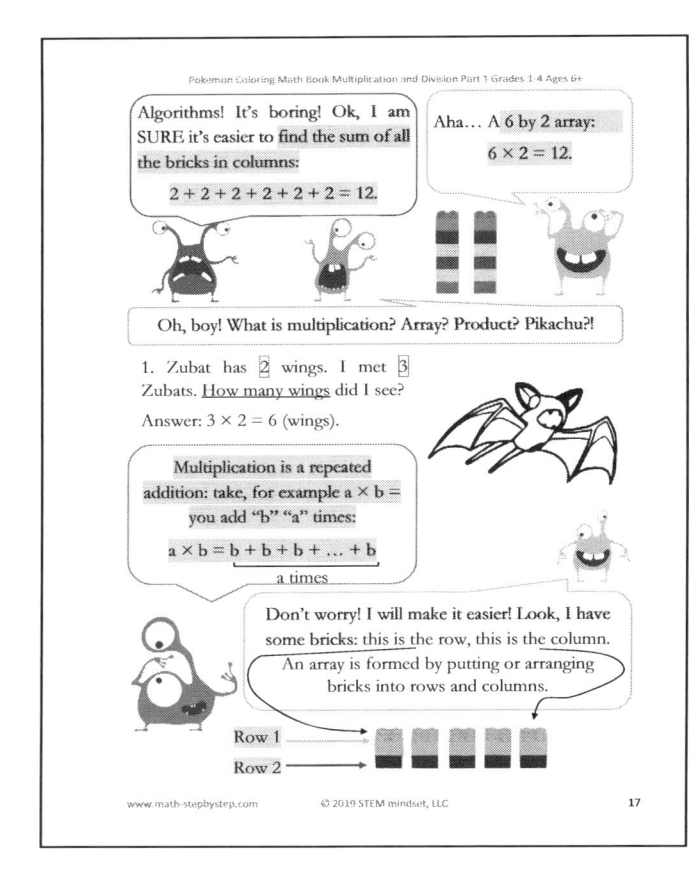

I divide 15 bricks into 5 equal parts. Hate math!

I will circle 15 bricks: each circle has only one (1) brick since I divide 15 by 15.

What's all the fuss about, Pickles? A fella's got to learn multiplication sometime...

$15 \div 15 = 1$

Aha... any number divided by itself equals 1:
$2 \div 2 = 1$; $85 \div 85 = 1$; $1 \div 1 = 1$.

Any number multiplied by 1 equals itself since we take only 1 group of this number.

Brilliant! Then, any number multiplied by 0 equals 0. Here, I take 0 groups of 11 bricks ☺.

$0 \times 11 = 0$

$1 \times 11 = 11$

Page 16

1. **Add or multiply.**

Take the bricks or cubes and arrange them so that each row has 6 bricks and each column has 2 bricks.
How many bricks do you have in all?
_____.

I know, I know! Find the sum of all the bricks in rows:
$6 + 6 = 12$.
Or I have
a 2 by 6 array of bricks:
$2 \times 6 = 12$.

| Groups or Rows | Bricks or Objects in each row | Total in all rows or all bricks |

Yeah, Sunny! Math is the only language that makes any sense to me!

Don't be silly! Math is a terrible experience!

I would multiply.
The product of 2 bricks per column and 5 bricks per row equals: $2 \times 5 = 10$
Then, we add 2 more bricks: $10 + 2 = 12$.

Page 17

Algorithms! It's boring! Ok, I am SURE it's easier to find the sum of all the bricks in columns:
$2 + 2 + 2 + 2 + 2 + 2 = 12$.

Aha... A 6 by 2 array:
$6 \times 2 = 12$.

Oh, boy! What is multiplication? Array? Product? Pikachu?!

1. Zubat has 2 wings. I met 3 Zubats. How many wings did I see?

Answer: $3 \times 2 = 6$ (wings).

Multiplication is a repeated addition: take, for example $a \times b$ = you add "b" "a" times:
$a \times b = \underbrace{b + b + b + \ldots + b}_{a \text{ times}}$

Don't worry! I will make it easier! Look, I have some bricks: this is the row, this is the column.
An array is formed by putting or arranging bricks into rows and columns.

Row 1 →
Row 2 →

Page 18

Aha... this is a 2 by 5 array. It has 2 rows and 5 columns. So, I count rows first, right?
And I write it as $2 \times 5 = 10$
If I make an addition number sentence, I need to add 5 bricks 2 times: $5 + 5 = 10$

2 rows

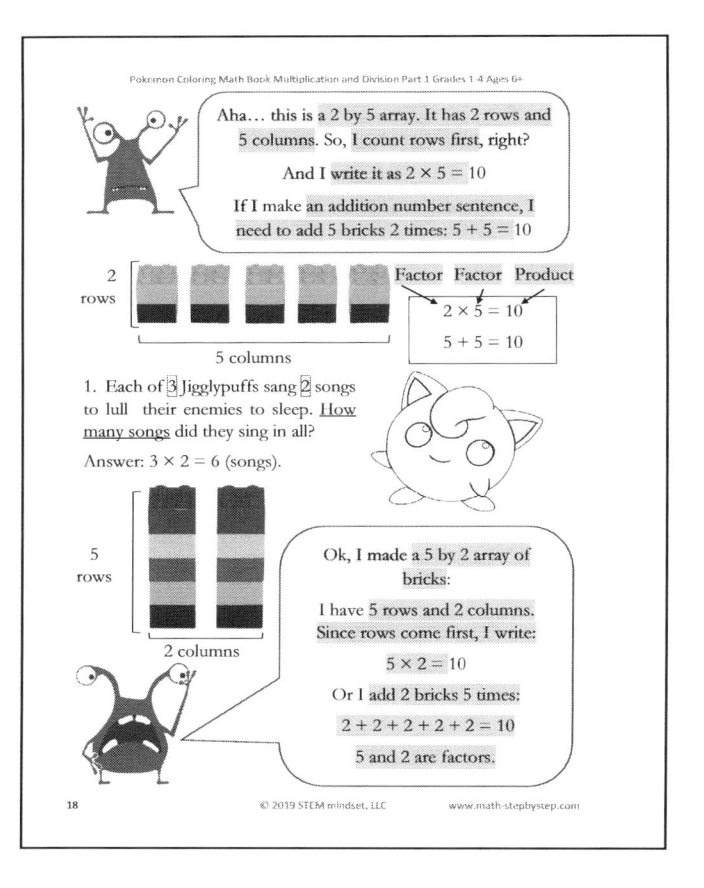

Factor Factor Product
$2 \times 5 = 10$
$5 + 5 = 10$

5 columns

1. Each of 3 Jigglypuffs sang 2 songs to lull their enemies to sleep. How many songs did they sing in all?

Answer: $3 \times 2 = 6$ (songs).

5 rows

2 columns

Ok, I made a 5 by 2 array of bricks:
I have 5 rows and 2 columns. Since rows come first, I write:
$5 \times 2 = 10$
Or I add 2 bricks 5 times:
$2 + 2 + 2 + 2 + 2 = 10$
5 and 2 are factors.

Page 19

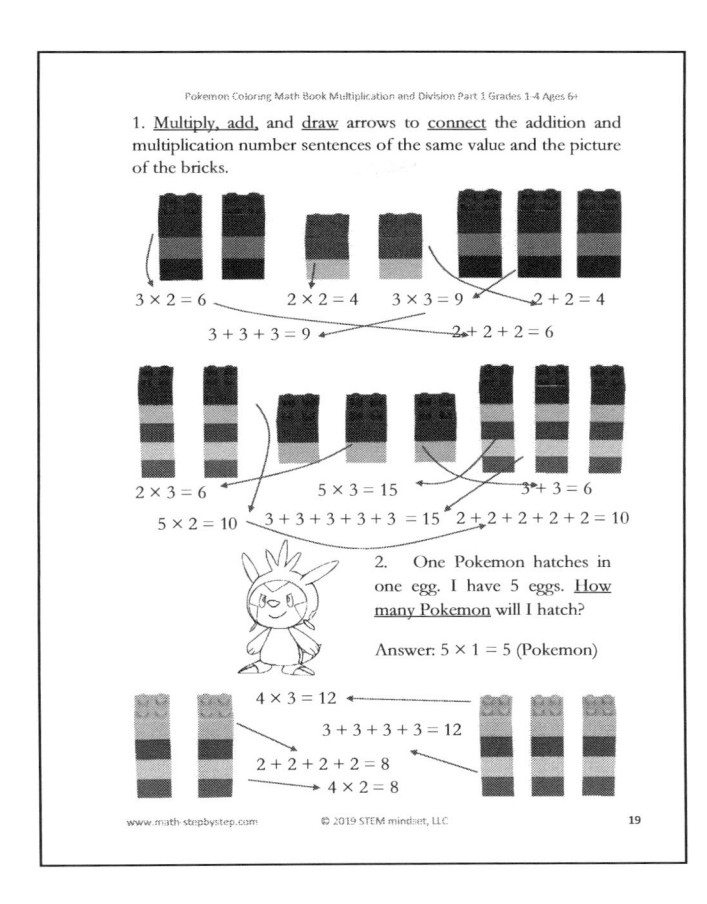

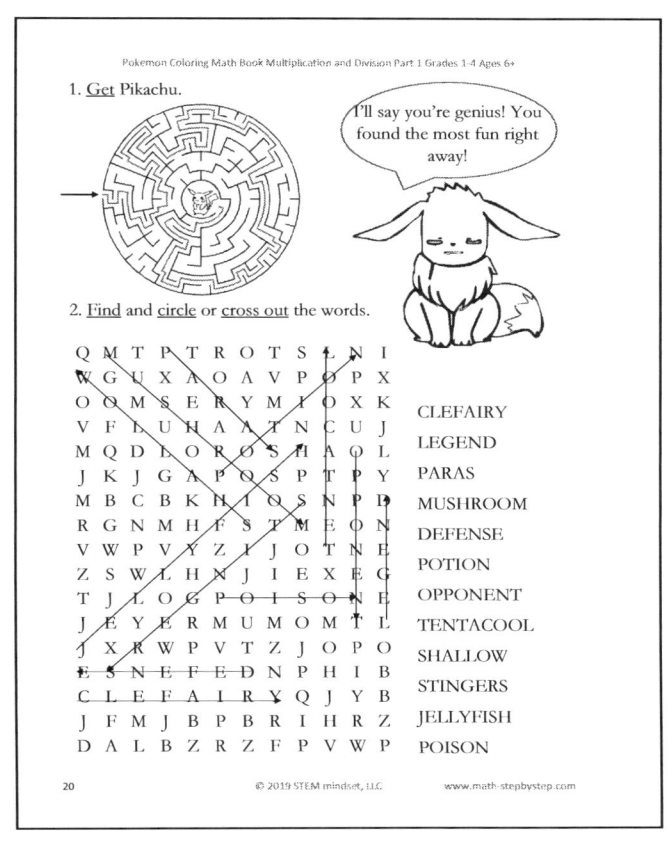

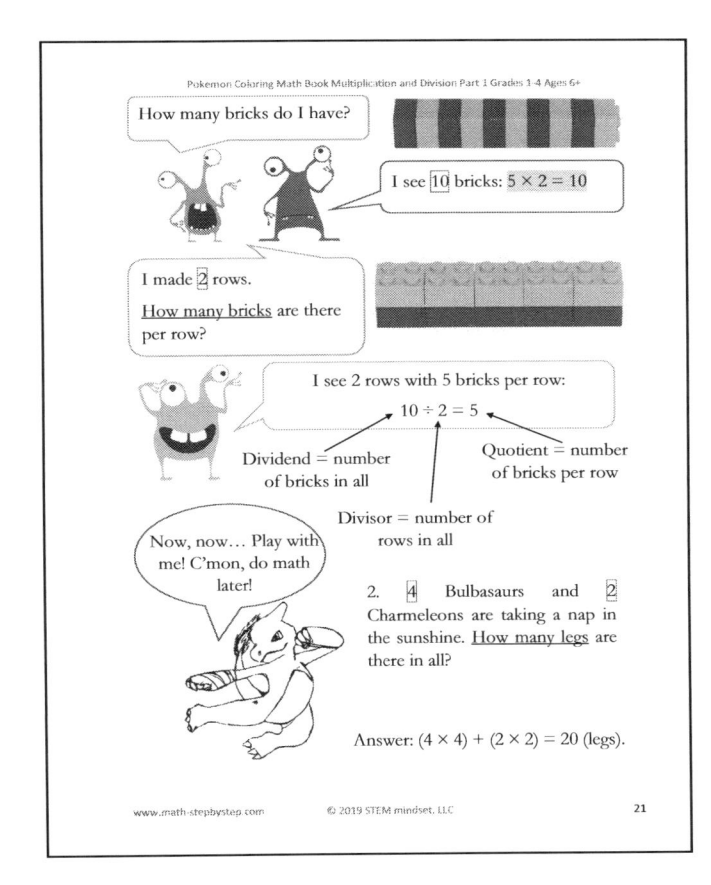

1. <u>Multiply</u>, <u>add</u>, and <u>draw</u> arrows to <u>connect</u> the addition and multiplication number sentences of the same value and the picture of the bricks.

$3 \times 2 = 6$ $2 \times 2 = 4$ $3 \times 3 = 9$ $2 + 2 = 4$

$3 + 3 + 3 = 9$ $2 + 2 + 2 = 6$

$2 \times 3 = 6$ $5 \times 3 = 15$ $3 + 3 = 6$

$5 \times 2 = 10$ $3 + 3 + 3 + 3 + 3 = 15$ $2 + 2 + 2 + 2 + 2 = 10$

2. One Pokemon hatches in one egg. I have 5 eggs. <u>How many Pokemon</u> will I hatch?

Answer: $5 \times 1 = 5$ (Pokemon)

$4 \times 3 = 12$
$3 + 3 + 3 = 12$
$2 + 2 + 2 + 2 = 8$
$4 \times 2 = 8$

Page 20

1. <u>Get</u> Pikachu.

I'll say you're genius! You found the most fun right away!

2. <u>Find</u> and <u>circle</u> or <u>cross out</u> the words.

```
Q M T P T R O T S L N I
W G U X A O A V P O P X
O O M S E R Y M I O X K      CLEFAIRY
V F L U H A A T N C U J      LEGEND
M Q D L O R O S H A O L      PARAS
J K J G A P Q S P T P Y      MUSHROOM
M B C B K H I O S N P B      DEFENSE
R G N M H F S T M E O N      POTION
V W P V Y Z I J O T N E      OPPONENT
Z S W L H N J I E X E G      TENTACOOL
T J L O G P O I S O N E      SHALLOW
J E Y E R M U M O M T L      STINGERS
J X R W P V T Z J O P O      JELLYFISH
E S N E F E D N P H I B      POISON
C L E F A I R Y Q J Y B
J F M J B P B R I H R Z
D A L B Z R Z F P V W P
```

CLEFAIRY
LEGEND
PARAS
MUSHROOM
DEFENSE
POTION
OPPONENT
TENTACOOL
SHALLOW
STINGERS
JELLYFISH
POISON

Page 21

How many bricks do I have?

I see $\boxed{10}$ bricks: $5 \times 2 = 10$

I made $\boxed{2}$ rows. <u>How many bricks</u> are there per row?

I see 2 rows with 5 bricks per row:
$10 \div 2 = 5$

Dividend = number of bricks in all
Quotient = number of bricks per row
Divisor = number of rows in all

Now, now... Play with me! C'mon, do math later!

2. $\boxed{4}$ Bulbasaurs and $\boxed{2}$ Charmeleons are taking a nap in the sunshine. <u>How many legs</u> are there in all?

Answer: $(4 \times 4) + (2 \times 2) = 20$ (legs).

Page 22

I made $\boxed{5}$ columns. <u>How many bricks</u> are there per column?

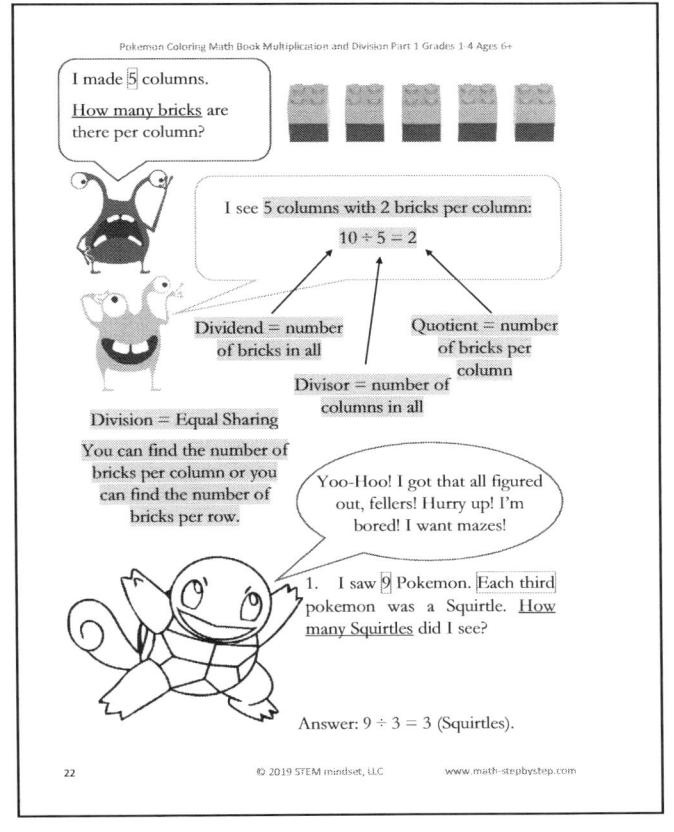

I see 5 columns with 2 bricks per column:
$10 \div 5 = 2$

Dividend = number of bricks in all
Quotient = number of bricks per column
Divisor = number of columns in all

Division = Equal Sharing

You can find the number of bricks per column or you can find the number of bricks per row.

Yoo-Hoo! I got that all figured out, fellers! Hurry up! I'm bored! I want mazes!

1. I saw $\boxed{9}$ Pokemon. Each third pokemon was a Squirtle. <u>How many Squirtles</u> did I see?

Answer: $9 \div 3 = 3$ (Squirtles).

Page 23

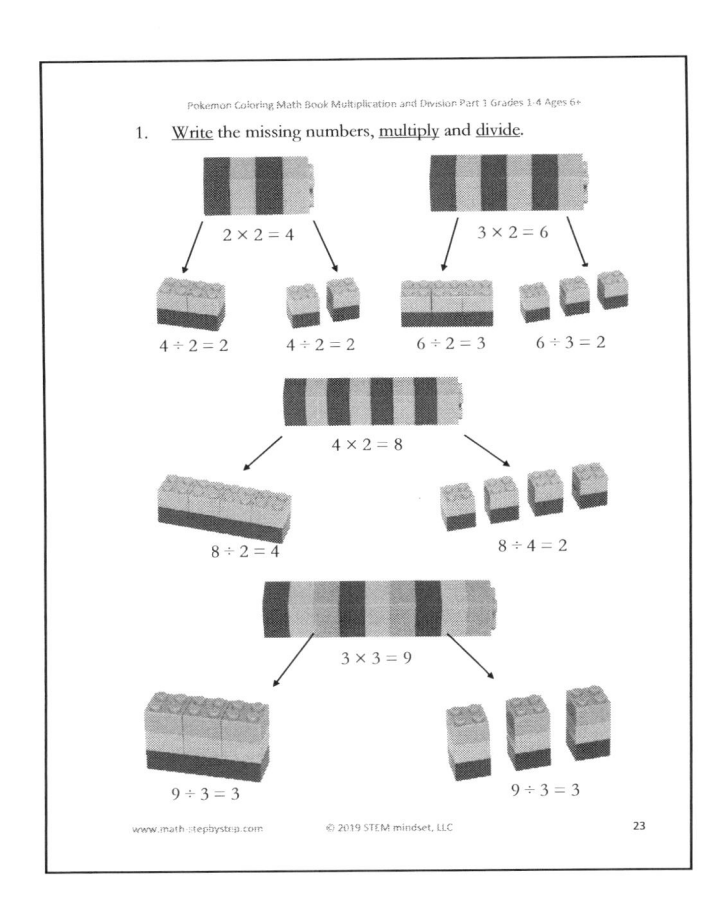

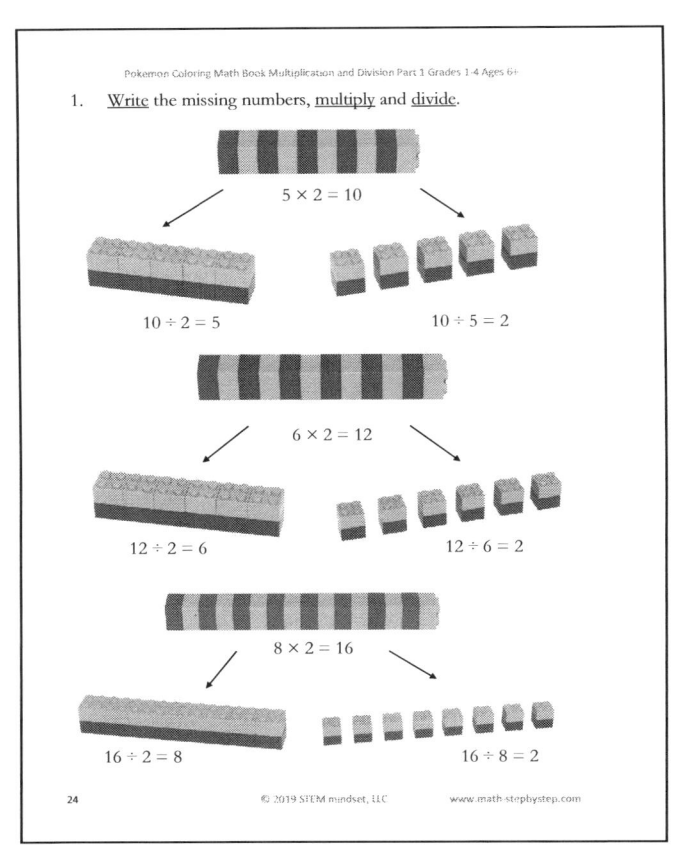

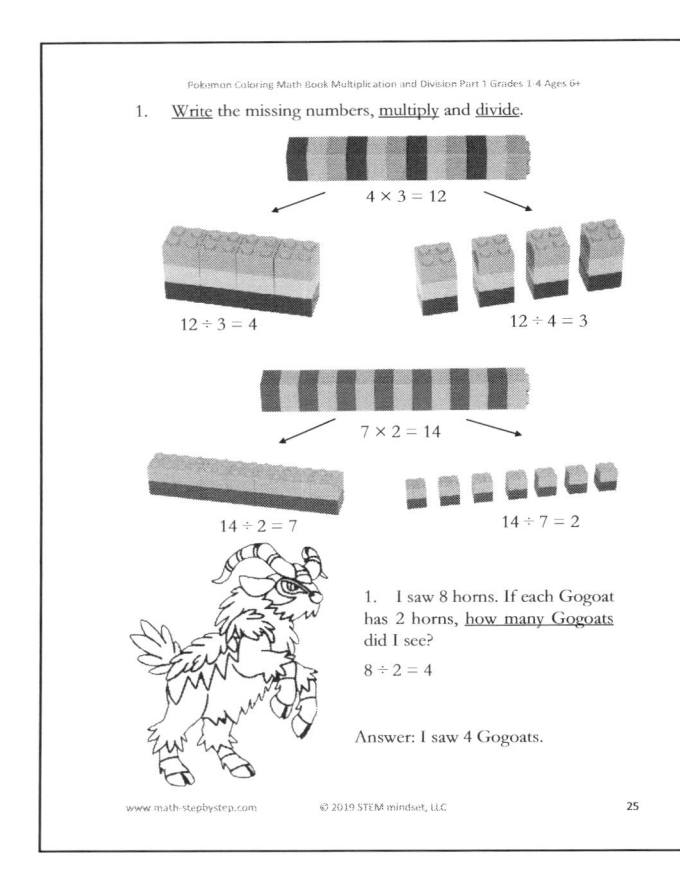

1. <u>Write</u> the missing numbers, <u>multiply</u> and <u>divide</u>.

$2 \times 2 = 4$ $3 \times 2 = 6$

$4 \div 2 = 2$ $4 \div 2 = 2$ $6 \div 2 = 3$ $6 \div 3 = 2$

$4 \times 2 = 8$

$8 \div 2 = 4$ $8 \div 4 = 2$

$3 \times 3 = 9$

$9 \div 3 = 3$ $9 \div 3 = 3$

Page 24

1. <u>Write</u> the missing numbers, <u>multiply</u> and <u>divide</u>.

$5 \times 2 = 10$

$10 \div 2 = 5$ $10 \div 5 = 2$

$6 \times 2 = 12$

$12 \div 2 = 6$ $12 \div 6 = 2$

$8 \times 2 = 16$

$16 \div 2 = 8$ $16 \div 8 = 2$

Page 25

1. <u>Write</u> the missing numbers, <u>multiply</u> and <u>divide</u>.

$4 \times 3 = 12$

$12 \div 3 = 4$ $12 \div 4 = 3$

$7 \times 2 = 14$

$14 \div 2 = 7$ $14 \div 7 = 2$

1. I saw 8 horns. If each Gogoat has 2 horns, <u>how many Gogoats</u> did I see?

$8 \div 2 = 4$

Answer: I saw 4 Gogoats.

Page 26

1. <u>Write</u> the missing numbers, <u>multiply</u> and <u>divide</u>.

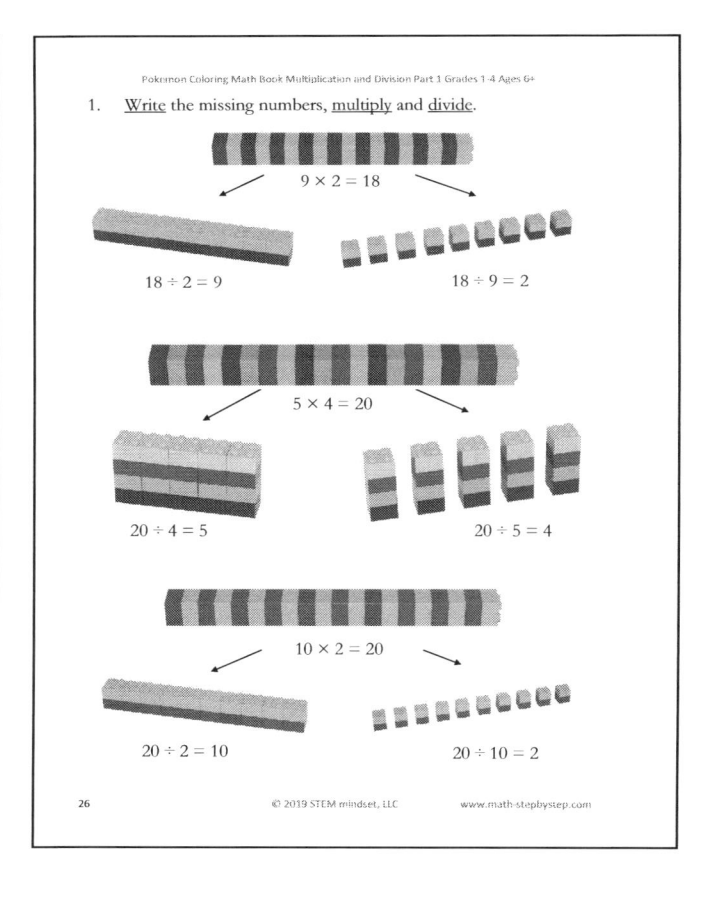

$9 \times 2 = 18$

$18 \div 2 = 9$ $18 \div 9 = 2$

$5 \times 4 = 20$

$20 \div 4 = 5$ $20 \div 5 = 4$

$10 \times 2 = 20$

$20 \div 2 = 10$ $20 \div 10 = 2$

Page 27

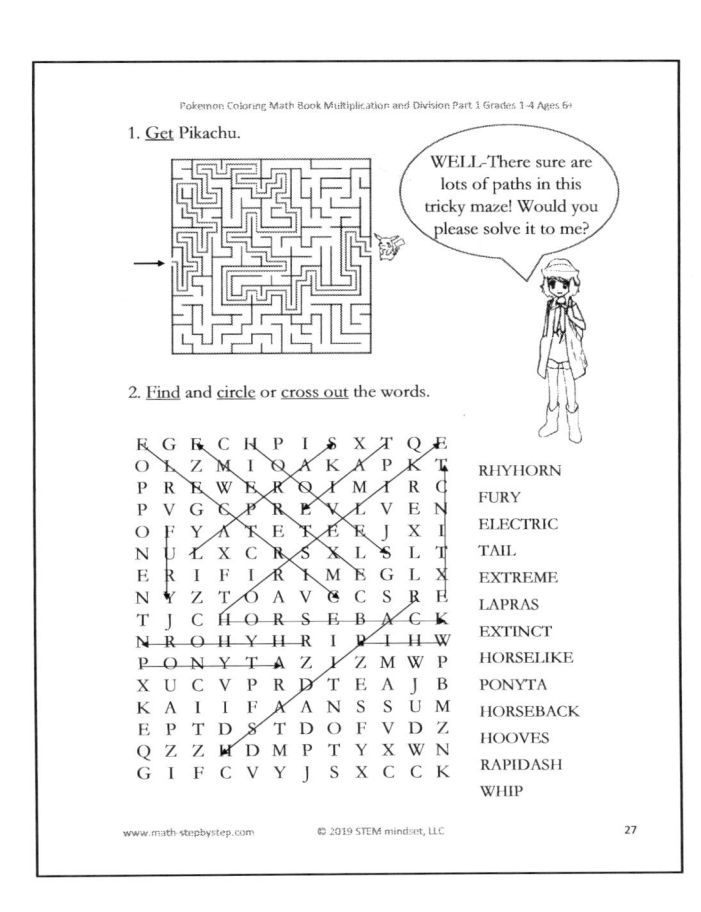

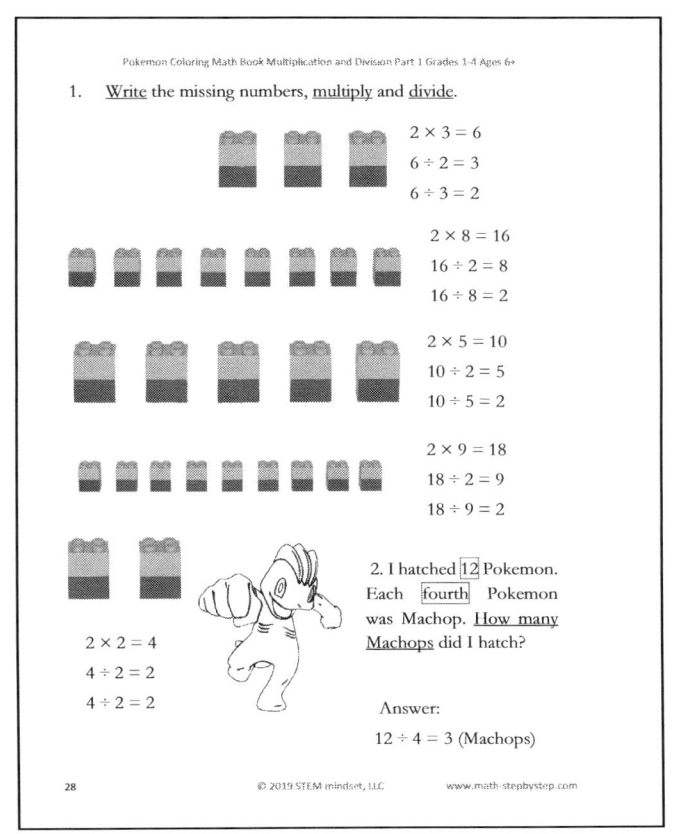

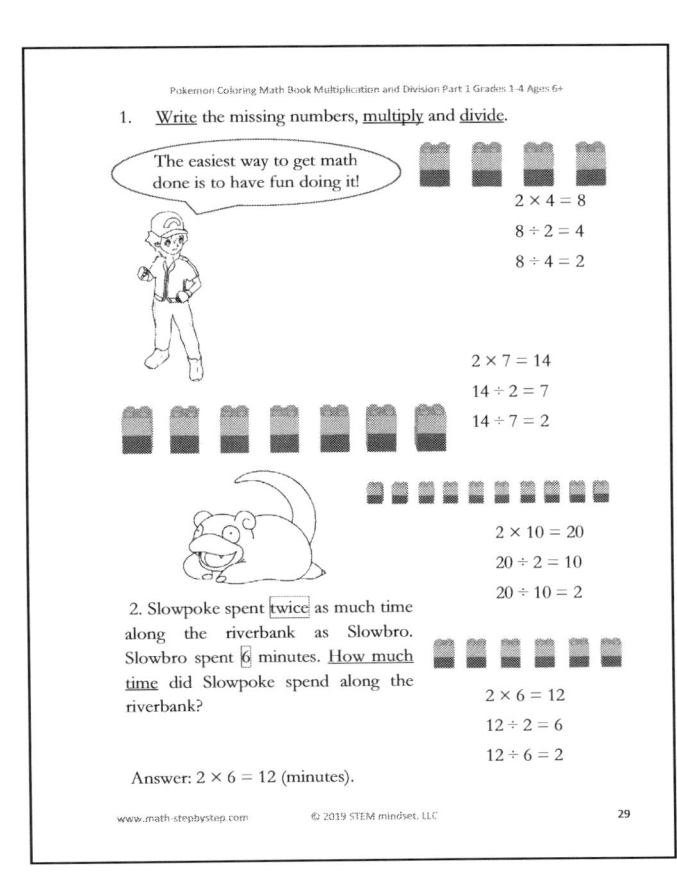

1. <u>Get</u> Pikachu.

WELL—There sure are lots of paths in this tricky maze! Would you please solve it to me?

2. <u>Find</u> and <u>circle</u> or <u>cross out</u> the words.

```
R G R C H P I S X T Q E
O L Z M I Q A K A P K T
P R E W E R O I M I R C
P V G C P R E V L V E N
O F Y A T E T E E J X I
N U L X C R S X L S L T
E R I F I R I M E G L X
N Y Z T O A V C C S R E
T J C H O R S E B A C K
N R O H Y H R I D I H W
P O N Y T A Z J Z M W P
X U C V P R D T E A J B
K A I I F A A N S S U M
E P T D S T D O F V D Z
Q Z Z H D M P T Y X W N
G I F C V Y J S X C C K
```

RHYHORN
FURY
ELECTRIC
TAIL
EXTREME
LAPRAS
EXTINCT
HORSELIKE
PONYTA
HORSEBACK
HOOVES
RAPIDASH
WHIP

Page 28

1. <u>Write</u> the missing numbers, <u>multiply</u> and <u>divide</u>.

$2 \times 3 = 6$
$6 \div 2 = 3$
$6 \div 3 = 2$

$2 \times 8 = 16$
$16 \div 2 = 8$
$16 \div 8 = 2$

$2 \times 5 = 10$
$10 \div 2 = 5$
$10 \div 5 = 2$

$2 \times 9 = 18$
$18 \div 2 = 9$
$18 \div 9 = 2$

$2 \times 2 = 4$
$4 \div 2 = 2$
$4 \div 2 = 2$

2. I hatched 12 Pokemon. Each fourth Pokemon was Machop. <u>How many Machops</u> did I hatch?

Answer:
$12 \div 4 = 3$ (Machops)

Page 29

1. <u>Write</u> the missing numbers, <u>multiply</u> and <u>divide</u>.

The easiest way to get math done is to have fun doing it!

$2 \times 4 = 8$
$8 \div 2 = 4$
$8 \div 4 = 2$

$2 \times 7 = 14$
$14 \div 2 = 7$
$14 \div 7 = 2$

$2 \times 10 = 20$
$20 \div 2 = 10$
$20 \div 10 = 2$

$2 \times 6 = 12$
$12 \div 2 = 6$
$12 \div 6 = 2$

2. Slowpoke spent twice as much time along the riverbank as Slowbro. Slowbro spent 6 minutes. <u>How much time</u> did Slowpoke spend along the riverbank?

Answer: $2 \times 6 = 12$ (minutes).

Page 30

1. <u>Multiply</u>.

```
    3        5        8        1        4
×   2    ×   2    ×   2    ×   2    ×   2
    6       10       16        2        8

    6        9        7        2       10
×   2    ×   2    ×   2    ×   2    ×   2
   12       18       14        4       20
```

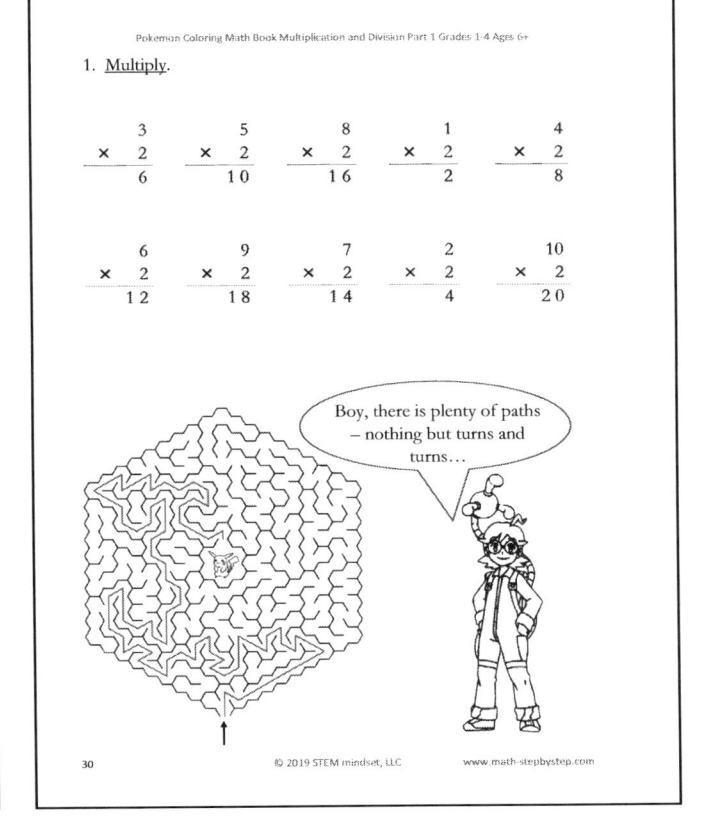

Boy, there is plenty of paths – nothing but turns and turns...

Page 31

1. Make 2 rows. X shows the number of columns. Find X. Write the missing numbers, multiply and divide.

$2 \times X = 10$ $X = 5$
$10 \div 5 = 2$ $10 \div 2 = 5$
2 rows
5 bricks in each row

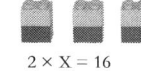

$2 \times X = 16$ $X = 8$
$16 \div 8 = 2$ $16 \div 2 = 8$
2 rows
8 bricks in each row

$2 \times X = 6$ $X = 3$
$6 \div 3 = 2$ $6 \div 2 = 3$
2 rows
3 bricks in each row

2. Wartortle gets 2 scratches for each battle. It had 7 battles. How many scratches does it have in all?

Answer: $7 \times 2 = 14$ (scratches).

Page 32

1. Make 2 rows. X shows the number of columns. Find X. Write the missing numbers, multiply and divide.

$2 \times X = 20$ $X = 10$
$20 \div 10 = 2$ $20 \div 2 = 10$
2 rows
10 bricks in each row

$2 \times X = 8$ $X = 4$
$8 \div 4 = 2$ $8 \div 2 = 4$
2 rows
4 bricks in each row

$2 \times X = 14$ $X = 7$
$14 \div 7 = 2$ $14 \div 2 = 7$
2 rows
7 bricks in each row

2. Blastoise fired 16 water bullets. Each second bullet hit the target. How many water bullets hit the target?

Answer: $16 \div 2 = 8$ (water bullets).

Page 33

1. Make 2 rows. X shows the number of columns. Find X. Write the missing numbers, multiply and divide.

$2 \times X = 12$ $X = 6$
$12 \div 6 = 2$ $12 \div 2 = 6$
2 rows
6 bricks in each row

$2 \times X = 4$ $X = 2$
$4 \div 2 = 2$ $4 \div 2 = 2$
2 rows
2 bricks in each row

$2 \times X = 18$ $X = 9$
$18 \div 9 = 2$ $18 \div 2 = 9$
2 rows
9 bricks in each row

2. Mankey had 5 fits of temper. Each fit was preceded by 3 violent tremors. How many tremors did it have in all?

Answer: $5 \times 3 = 15$ (tremors).

Page 34

1. Write the missing numbers.

	2	2	10	2	2
×	4	2	2	6	9
	8	4	20	12	18

	2	2	2	2	2
×	5	8	3	7	1
	10	16	6	14	2

"Perhaps this time I will get this little monster!"

Page 35

1. <u>Add</u> and <u>multiply</u>.

$2 + 2 + 2 = 6$	$3 \times 2 = 6$
$2 + 2 + 2 + 2 = 8$	$4 \times 2 = 8$
$2 + 2 + 2 + 2 + 2 = 10$	$5 \times 2 = 10$
$2 + 2 + 2 + 2 + 2 + 2 = 12$	$6 \times 2 = 12$
$2 + 2 + 2 + 2 + 2 + 2 + 2 = 14$	$7 \times 2 = 14$
$2 + 2 + 2 + 2 + 2 + 2 + 2 + 2 = 16$	$8 \times 2 = 16$
$2 + 2 + 2 + 2 + 2 + 2 + 2 + 2 + 2 = 18$	$9 \times 2 = 18$
$2 + 2 + 2 + 2 + 2 + 2 + 2 + 2 + 2 + 2 = 20$	$10 \times 2 = 20$

2. <u>Take</u> 18 bricks to answer the questions.

<u>Make</u> 2 rows.

<u>How many</u> columns can you make with 12 bricks?   6

<u>How many</u> columns can you make with the leftover bricks?   3

<u>How many</u> columns can you make out of all the bricks?   9.

Page 36

1. <u>Get</u> Pikachu.

"This word search looks like an adventure! It has adjectives! W-O-O-S-H!"

2. <u>Find</u> and <u>circle</u> or <u>cross out</u> the words.

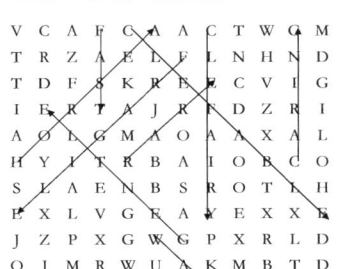

CLEFABLE
CLEFAIRY
GENTLE
CARING
TECHNIQUES
FAST
SPEED
RARE
HORSEA
FRAGILE
SURFACE
WATER

Page 37

1. <u>Write</u> the missing numbers, <u>multiply</u> and <u>divide</u>.

$3 \times 3 = 3$
$9 \div 3 = 3$
$9 \div 3 = 3$

$3 \times 2 = 6$
$6 \div 3 = 2$
$6 \div 2 = 3$

$3 \times 8 = 24$ $24 \div 3 = 8$ $24 \div 8 = 3$

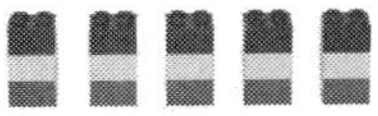

$3 \times 5 = 15$
$15 \div 3 = 5$
$15 \div 5 = 3$

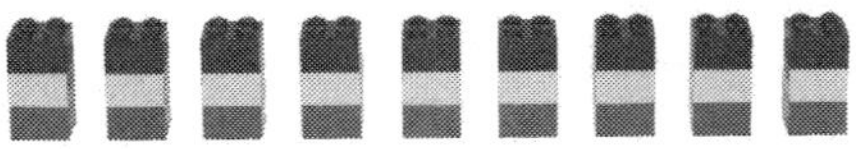

$3 \times 9 = 27$ $27 \div 3 = 9$ $27 \div 9 = 3$

2. Rhyhorn thinks about demolishing each of its target 3 times a day. If it has 4 targets, <u>how many</u> <u>thoughts</u> does it have daily?

Answer: $4 \times 3 = 12$ (thoughts).

Page 38

1. <u>Write</u> the missing numbers, <u>multiply</u> and <u>divide</u>.

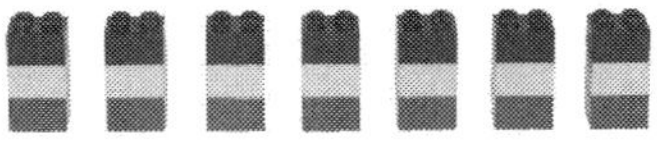

$3 \times 7 = 21$
$21 \div 3 = 7$
$21 \div 7 = 3$

$3 \times 4 = 12$
$12 \div 3 = 4$
$12 \div 4 = 3$

$3 \times 10 = 30$ $30 \div 3 = 10$ $30 \div 10 = 3$

$3 \times 6 = 18$
$18 \div 3 = 6$
$18 \div 6 = 3$

2. Jolteon had 3 battles. Each battle it called down 4 thunderbolts! <u>How many thunderbolts</u> did it call down in all?

Answer: $3 \times 4 = 12$ (thunderbolts).

Page 39

1. <u>Multiply</u>.

	3	5	8	1	4
×	3	3	3	3	3
	9	15	24	3	12

	6	9	7	2	10
×	3	3	3	3	3
	18	27	21	6	30

Look! There Pikachu is! What are you waiting for? GET IT!

Page 40

1. <u>Make</u> $\boxed{3}$ rows. $\boxed{X}$ shows the number of columns. <u>Find</u> X. <u>Write</u> the missing numbers, <u>multiply</u> and <u>divide</u>.

$3 \times X = 30$ $X = 10$
$30 \div 10 = 3$ $30 \div 3 = 10$
3 rows   10 bricks in each row

$3 \times X = 18$ $X = 6$
$18 \div 6 = 3$ $18 \div 3 = 6$
3 rows
6 bricks in each row

$3 \times X = 27$ $X = 9$
$27 \div 9 = 3$ $27 \div 3 = 9$
3 rows   9 bricks in each row

2. Moltres found $\boxed{3}$ volcanoes. Each time he bathed $\boxed{5}$ times in lava. <u>How many times</u> did he bathe in all?

Answer: $3 \times 5 = 15$ (times).

Page 41

1. <u>Make</u> $\boxed{3}$ rows. $\boxed{X}$ shows the number of columns. <u>Find</u> X. <u>Write</u> the missing numbers, <u>multiply</u> and <u>divide</u>.

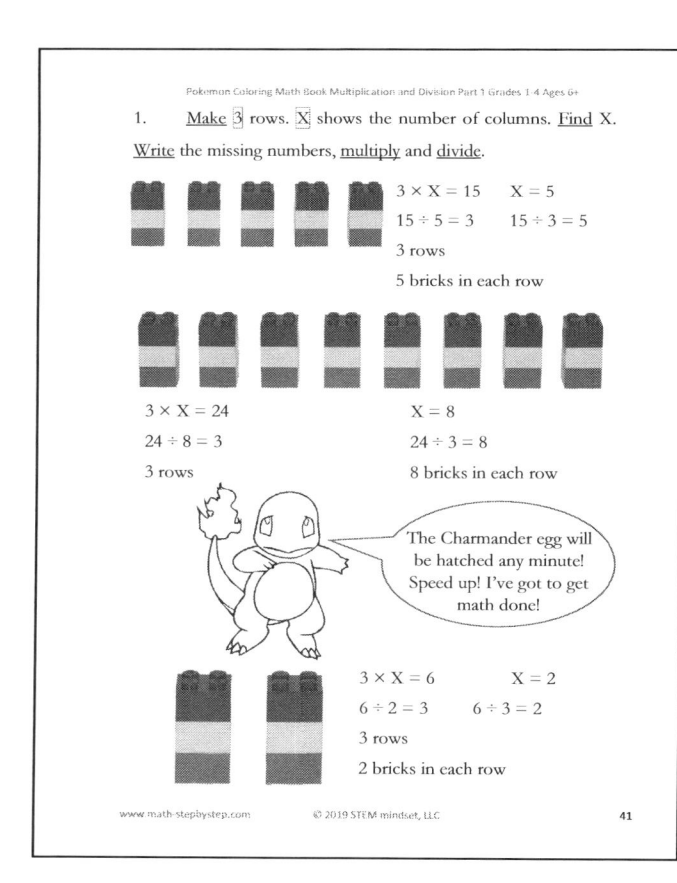

$3 \times X = 15$ $X = 5$
$15 \div 5 = 3$ $15 \div 3 = 5$
3 rows
5 bricks in each row

$3 \times X = 24$ $X = 8$
$24 \div 8 = 3$ $24 \div 3 = 8$
3 rows   8 bricks in each row

The Charmander egg will be hatched any minute! Speed up! I've got to get math done!

$3 \times X = 6$ $X = 2$
$6 \div 3 = 3$ $6 \div 3 = 2$
3 rows
2 bricks in each row

Page 42

1. <u>Make</u> $\boxed{3}$ rows. $\boxed{X}$ shows the number of columns. <u>Find</u> X. <u>Write</u> the missing numbers, <u>multiply</u> and <u>divide</u>.

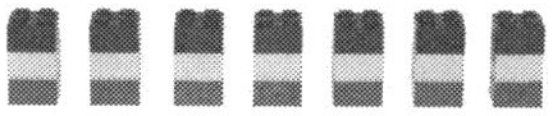

$3 \times X = 21$ $X = 7$
$21 \div 7 = 3$ $21 \div 3 = 7$
3 rows
7 bricks in each row

$3 \times X = 9$ $X = 3$
$9 \div 3 = 3$ $9 \div 3 = 3$
3 rows   3 bricks in each row

$3 \times X = 12$ $X = 4$
$12 \div 4 = 3$ $12 \div 3 = 4$
3 rows
4 bricks in each row

2. Moltres bathed $\boxed{24}$ times in lava. If he bathed $\boxed{8}$ times in each volcano, <u>how many volcanoes</u> did he dive into?

Answer: $24 \div 8 = 3$ (volcanoes).

Page 43

1. <u>Write</u> the missing numbers.

```
    3        1        7        4        6
×   3    ×   3    ×   3    ×   3    ×   3
    9        3       21       12       18

    5        9        2        8       10
×   3    ×   3    ×   3    ×   3    ×   3
   15       27        6       24       30
```

Now, with such a Smarty and such a Sunny, how can we fail?

Page 44

1. <u>Add</u> and <u>multiply</u>. The easiest way to count 3's is counting by groups of two 3's or of four 3's. Two 3's (3 + 3) equal 6. Or four 3's (3 + 3 + 3 + 3) equal 12.

3 + 3 = 6	2 × 3 = 6
3 + 3 + 3 = 6 + 3 = 9	3 × 3 = 9
3 + 3 + 3 + 3 = 12	4 × 3 = 12
3 + 3 + 3 + 3 + 3 = 15	5 × 3 = 15
3 + 3 + 3 + 3 + 3 + 3 = 18	6 × 3 = 18
3 + 3 + 3 + 3 + 3 + 3 + 3 = 21	7 × 3 = 21
3 + 3 + 3 + 3 + 3 + 3 + 3 + 3 = 24	8 × 3 = 24
3 + 3 + 3 + 3 + 3 + 3 + 3 + 3 + 3 = 27	9 × 3 = 27
3 + 3 + 3 + 3 + 3 + 3 + 3 + 3 + 3 + 3 = 30	10 × 3 = 30

2. <u>Take</u> 27 bricks to answer the questions.

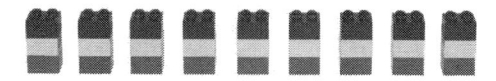

<u>Make</u> 3 rows.

<u>How many</u> columns can you make with 15 bricks?   5

<u>How many</u> columns can you make with the leftover bricks?   4

<u>How many</u> columns can you make out of all the bricks?   9.

Page 45

1. <u>Get</u> Pikachu.

Pickles, you're mking me nervous! I wish I'd be hatched into something BIG!

2. <u>Find</u> and <u>circle</u> or <u>cross out</u> the words.

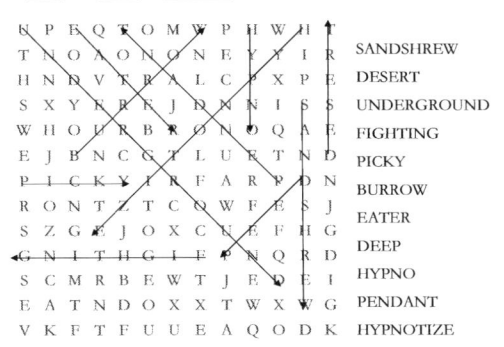

SANDSHREW
DESERT
UNDERGROUND
FIGHTING
PICKY
BURROW
EATER
DEEP
HYPNO
PENDANT
HYPNOTIZE

Page 46

1. <u>Multiply</u> and <u>divide</u>.

2 × 4 = 8	2 × ❀ = 8	❀ = 4	8 ÷ 2 = 4
2 × 2 = 4	2 × ❀ = 4	❀ = 2	4 ÷ 2 = 2
2 × 5 = 10	2 × ❀ = 10	❀ = 5	10 ÷ 2 = 5
2 × 3 = 6	2 × ❀ = 6	❀ = 3	6 ÷ 2 = 3
3 × 3 = 9	3 × ❀ = 9	❀ = 3	9 ÷ 3 = 3
3 × 2 = 6	3 × ❀ = 6	❀ = 2	6 ÷ 3 = 2
4 × 2 = 8	4 × ❀ = 8	❀ = 2	8 ÷ 4 = 2
5 × 2 = 10	5 × X = 10	X = 2	10 ÷ 5 = 2

2. <u>Divide</u> (find <u>how many</u> groups of 2, 3, 4, or 5 you can make out of all the bricks) and <u>write down</u> the remainder in the parentheses.

Remainder is what is left over after the division operation.

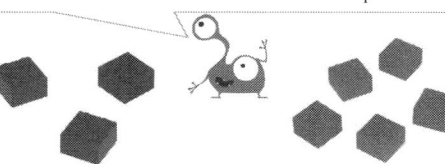

3 ÷ 2 = 1 (1)   5 ÷ 2 = 2 (1)
                                    5 ÷ 5 = 1 (0)
3 ÷ 3 = 1 (0)   5 ÷ 4 = 1 (1)
                                    5 ÷ 3 = 1 (2)

Pokemon Coloring Math Book Multiplication and Division Part 1 Grades 1-4 Ages 6+

1. <u>Divide</u> and <u>write</u> the remainder in the parentheses.

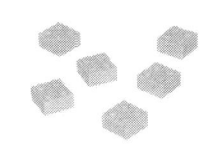

$6 ÷ 5 = 1\ (1)$
$6 ÷ 3 = 2\ (0)$
$6 ÷ 6 = 0\ (0)$
$6 ÷ 2 = 3\ (0)$
$6 ÷ 4 = 1\ (2)$

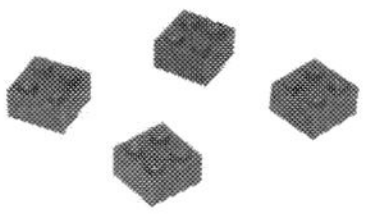

$4 ÷ 2 = 2\ (0)$
$4 ÷ 3 = 1\ (1)$
$4 ÷ 4 = 1\ (0)$

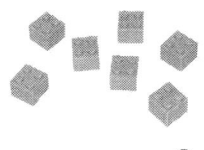

$7 ÷ 2 = 3\ (1)$
$7 ÷ 5 = 1\ (2)$
$7 ÷ 6 = 1\ (1)$
$7 ÷ 4 = 1\ (3)$
$7 ÷ 3 = 2\ (1)$
$7 ÷ 7 = 1\ (0)$

2. Mew had 6 battles. Each battle he turned invisible 4 times. <u>How many times</u> was he invisible in all?

Answer: $6 × 4 = 24$ (times).

Pokemon Coloring Math Book Multiplication and Division Part 1 Grades 1-4 Ages 6+

1. <u>Multiply</u> and <u>divide</u>.

$8 ÷ 2 = 4$
$18 ÷ 2 = 9$
$2 = 4$
$6 = 12$
$8 = 16$
$12 ÷ 2 = 6$
$10 ÷ 2 = 5$
$3 = 6$
$2 ×$
$5 = 10$
$14 ÷ 2 = 7$
$16 ÷ 2 = 8$
$7 = 14$
$9 = 18$
$6 ÷ 2 = 3$
$4 = 8$
$4 ÷ 2 = 2$

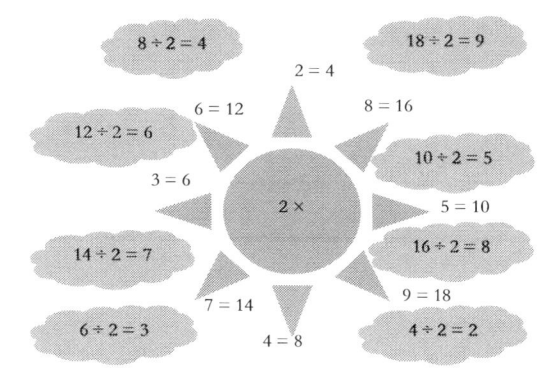

2. <u>Find</u> the missing divisor.

$15 ÷ X = 3$ $18 ÷ X = 6$ $12 ÷ X = 4$
$X = 5$ $X = 3$ $X = 3$
$9 ÷ ? = 3$ $8 ÷ ? = 2$ $6 ÷ ? = 2$
$? = 3$ $? = 4$ $? = 3$
$12 ÷ 🐭 = 2$ $16 ÷ 🐻 = 4$ $20 ÷ 🐗 = 5$
$🐭 = 6$ $🐻 = 4$ $🐗 = 4$

Pokemon Coloring Math Book Multiplication and Division Part 1 Grades 1-4 Ages 6+

1. <u>Divide</u> and <u>write</u> the remainder in the parentheses.

$9 ÷ 5 = 1\ (4)$
$9 ÷ 2 = 4\ (1)$
$9 ÷ 7 = 1\ (2)$
$9 ÷ 4 = 2\ (1)$
$9 ÷ 6 = 1\ (3)$
$9 ÷ 3 = 3\ (0)$

$8 ÷ 2 = 4\ (0)$
$8 ÷ 8 = 1\ (0)$
$8 ÷ 6 = 1\ (2)$
$8 ÷ 5 = 1\ (3)$
$8 ÷ 3 = 2\ (2)$
$8 ÷ 4 = 2\ (0)$
$8 ÷ 7 = 1\ (1)$

$10 ÷ 5 = 2\ (0)$ $10 ÷ 6 = 1\ (4)$
$10 ÷ 2 = 5\ (0)$ $10 ÷ 8 = 1\ (2)$
$10 ÷ 7 = 1\ (3)$ $10 ÷ 4 = 2\ (2)$
$10 ÷ 3 = 3\ (1)$ $10 ÷ 9 = 1\ (1)$

Pokemon Coloring Math Book Multiplication and Division Part 1 Grades 1-4 Ages 6+

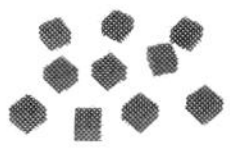

1. <u>Write</u> the missing numbers 2, 2, 2, 5, and 10 to make the equation true.

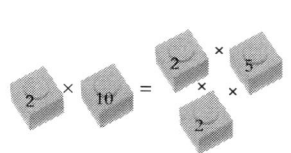

$2 × 10 = 2 × 5 × 2 × 2$

2. <u>Find</u> the missing factor or dividend.

$X × 2 = 18$ $X × 7 = 14$ $X × 4 = 12$
$X = 9$ $X = 2$ $X = 3$
$? × 5 = 15$ $? × 5 = 20$ $? × 2 = 12$
$? = 3$ $? = 4$ $? = 6$
$🐭 × 6 = 18$ $🐻 × 4 = 16$ $🐗 × 2 = 16$
$🐭 = 3$ $🐻 = 4$ $🐗 = 8$
$X ÷ 3 = 4$ $X ÷ 3 = 6$ $X ÷ 9 = 2$
$X = 12$ $X = 18$ $X = 18$
$? ÷ 6 = 2$ $? ÷ 4 = 5$ $? ÷ 3 = 5$
$? = 12$ $? = 20$ $? = 15$
$🐭 ÷ 10 = 2$ $🐻 ÷ 8 = 1$ $🐗 ÷ 1 = 4$
$🐭 = 20$ $🐻 = 8$ $🐗 = 4$

Pokemon Coloring Math Book Multiplication and Division Part 1 Grades 1-4 Ages 6+

1. <u>Get</u> Pikachu.

Here's where I get even for all the Team Rocket that ever pestered me.

2. <u>Find</u> and <u>circle</u> or <u>cross out</u> the words.

ZUBAT
TUNNELS
CAVES
ULTRASONIC
RADAR
GOLBAT
SCREECH
FANGS
BITE
CONFUSE
ENERGY

51

Pokemon Coloring Math Book Multiplication and Division Part 1 Grades 1-4 Ages 6+

1. <u>Write</u> the missing numbers 2, 2, 2, 2, 3, 3, and 4 to make the equation true.

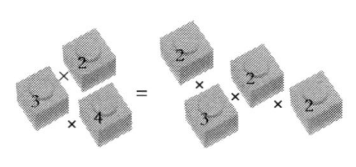

2. <u>Find</u> the missing factor or dividend.

X × 3 = 12	X × 3 = 18	X × 9 = 18
X = 4	X = 6	X = 2
? × 6 = 12	? × 4 = 20	? × 3 = 15
? = 2	? = 5	? = 5
🐭 × 10 = 20	🐱 × 8 = 8	🐀 × 1 = 4
🐭 = 2	🐱 = 1	🐀 = 4
X ÷ 2 = 9	X ÷ 7 = 2	X ÷ 4 = 3
X = 18	X = 14	X = 12
? ÷ 5 = 3	? ÷ 5 = 4	? ÷ 2 = 6
? = 15	? = 20	? = 12
🐭 ÷ 6 = 3	🐱 ÷ 4 = 4	🐀 ÷ 2 = 8
🐭 = 18	🐱 = 16	🐀 = 16

52

Pokemon Coloring Math Book Multiplication and Division Part 1 Grades 1-4 Ages 6+

1. <u>Compare</u> the number sentences, using ">", "<", or "=".

4 + 4 + 4	<	5 × 4		2 + 2 + 2 + 2	>	2 × 2
3 + 3	<	4 × 3		5 + 5 + 5	=	3 × 5
6 + 6	<	3 × 6		2 + 2 + 2	<	4 × 2
3 + 3 + 3 + 3	<	5 × 3		7 + 7	=	2 × 7
8 + 8	>	1 × 8		3 + 3 + 3	>	2 × 3

2. <u>Evaluate</u> each expression. <u>Indicate</u> the order of operations.

Are you kidding? First, we multiply in the parentheses (3 × 3 = 9), then, we add or subtract: 9 + 15 = 24.

3 × 5 = 15   (3 × 3) $\overset{12}{+}$ 15 = 24   (3 × 6) $\overset{12}{-}$ 12 = 6
3 × 2 = 6   (3 × 4) + 8 = 20   (3 × 10) − 15 = 15
3 × 6 = 18   (3 × 7) + 19 = 40   (3 × 4) − 6 = 6
3 × 9 = 27   (3 × 8) + 7 = 31   (3 × 5) − 7 = 8

2. I captured 3 Pidgeys om Monday. I captured four times as many Pidgeys on Friday as on Monday. <u>How many Pidgeys</u> did I capture in all?

Answer: (4 × 3) + 3 = 15 (Pidgeys).

53

Pokemon Coloring Math Book Multiplication and Division Part 1 Grades 1-4 Ages 6+

1. <u>Write</u> the missing numbers 2, 2, 2, 2, 4, 4, and 4 to make the equation true.

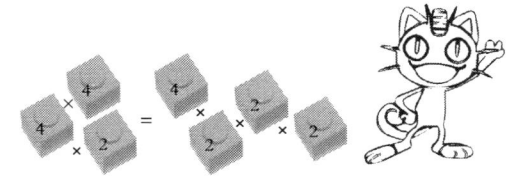

2. <u>Evaluate</u> each expression. <u>Indicate</u> the order of operations.

$\overset{1234}{(6 \times 2) - 5 + 10 - 8} = 12 - 5 + 10 - 8 = 7 + 10 - 8 = 17 - 8 = 9$

$\overset{1234}{(4 \times 4) - 3 + 5 - 9} = \underline{9}$

$\overset{1234}{(3 \times 5) - 8 + 4 - 6} = \underline{5}$

$\overset{1234}{(5 \times 4) - 6 + 3 - 8} = \underline{9}$

$\overset{1234}{(5 \times 2) - 4 + 12 - 3} = \underline{15}$

$\overset{1234}{(4 \times 2) - 1 + 13 - 6} = \underline{14}$

$\overset{1234}{(3 \times 6) - 13 + 7 - 4} = \underline{8}$

$\overset{1234}{(3 \times 4) - 6 + 11 - 4} = \underline{13}$

54

1. <u>Multiply</u> and <u>divide</u>. <u>Connect</u> the three multiplication and division diagrams of the same value. The first one is done for you.

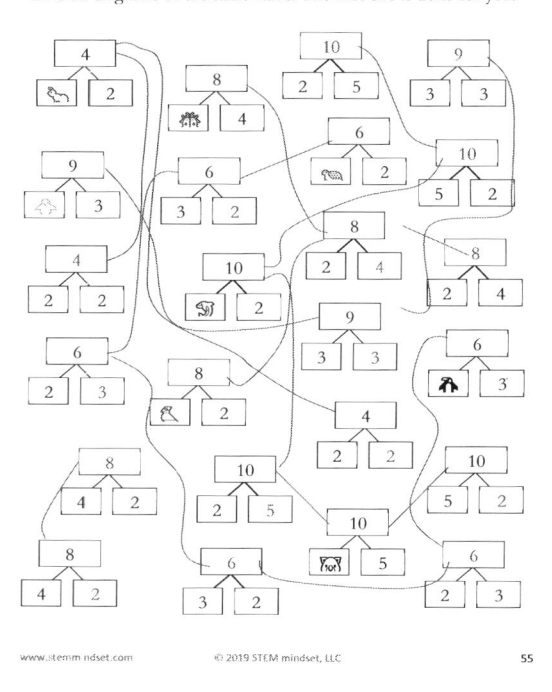

---

1. <u>Write</u> the missing numbers 2, 2, 3, 3, 3, 3, and 4 to make the equation true.

2. <u>Find</u> the missing divisor.

$18 \div X = 2$ $15 \div X = 5$ $18 \div X = 3$

$X = 9$ $X = 3$ $X = 6$

$20 \div ? = 4$ $16 \div ? = 8$ $8 \div ? = 4$

$? = 5$ $? = 2$ $? = 2$

$18 \div 🐾 = 9$ $16 \div 🐰 = 2$ $14 \div 🐗 = 7$

$🐾 = 2$ $🐰 = 8$ $🐗 = 2$

3. <u>Evaluate</u> each expression. <u>Indicate</u> the order of operations.

$(3 \times 6) - 15 + 17 - 4 = 16$ $(3 \times 4) - 11 + 19 - 8 = 12$

$(7 \times 2) - 12 + 14 - 8 = 8$ $(2 \times 9) - 17 + 14 - 7 = 8$

$(6 \times 3) - 13 + 10 - 6 = 9$ $(2 \times 8) - 12 + 9 - 11 = 2$

4. Pidgeotto used 6 times as many claws as pecks on his intruder. If it fiercely used 5 peckes on his intruder, <u>how many times</u> did it use its claws?

Answer: $5 \times 6 = 30$ (times).

---

1. <u>Multiply</u> and <u>divide</u>.

$2 \times 8 = 16$ $2 \times ❋ = 16$ $❋ = 8$ $16 \div 2 = 8$

$2 \times 6 = 12$ $2 \times ❋ = 12$ $❋ = 6$ $12 \div 2 = 6$

$2 \times 10 = 20$ $2 \times ❋ = 20$ $❋ = 10$ $20 \div 2 = 10$

$2 \times 7 = 14$ $2 \times ❋ = 14$ $❋ = 7$ $14 \div 2 = 7$

$2 \times 9 = 18$ $2 \times ❋ = 18$ $❋ = 9$ $18 \div 2 = 9$

$3 \times 5 = 15$ $3 \times ❋ = 15$ $❋ = 5$ $15 \div 3 = 5$

$3 \times 4 = 12$ $3 \times ❋ = 12$ $❋ = 4$ $12 \div 3 = 4$

$3 \times 6 = 18$ $3 \times ❋ = 18$ $❋ = 6$ $18 \div 3 = 6$

$4 \times 5 = 20$ $4 \times ❋ = 20$ $❋ = 5$ $20 \div 4 = 5$

$4 \times 4 = 16$ $4 \times ❋ = 16$ $❋ = 4$ $16 \div 4 = 4$

2. <u>Divide</u> and <u>write</u> the <u>remainder</u> in the parentheses.

$13 \div 5 = 2 \ (3)$ $13 \div 2 = 6 \ (1)$

$13 \div 7 = 1 \ (6)$ $13 \div 3 = 4 \ (1)$ $13 \div 8 = 1 \ (5)$

$13 \div 6 = 2 \ (1)$ $13 \div 9 = 1 \ (4)$ $13 \div 4 = 3 \ (1)$

---

1. <u>Get</u> Pikachu.

> Outa my way, Pikachu! I'm gonna roll up my sleeves and give you a good HUNT!

2. <u>Find</u> and <u>circle</u> or <u>cross out</u> the words.

```
E I Z R R A C S T W R W
A V R J E A F O O O H
Q E I G P L E O E O J I
P H D T A B K D G D W R
S I E P C E E A D S P L
P R S T A E D N I F I W
K D R I B Y T R P L G I
S S P I D G E O T T O N
Y T I L I G A T R Y V P
W I N G S Y H Q W P Z D
D D W Q R A B M X E I V
N L O J P H Z Q Q E R R
```

PIDGEY
CAPTURE
BIRD
TORNADOES
WINGS
WOODS
FLAPS
AGILITY
PIDGEOTTO
PIDGEOT
PROTECTIVE
WHIRLWIND

Page 59

1. <u>Multiply</u> and <u>divide</u>.

5 × 3 = 15	5 × ★ = 15 ★ = 3	15 ÷ 5 = 3
5 × 4 = 20	5 × ★ = 20 ★ = 4	20 ÷ 5 = 4
6 × 2 = 12	6 × ★ = 12 ★ = 2	12 ÷ 6 = 2
6 × 3 = 18	6 × ★ = 18 ★ = 3	18 ÷ 6 = 3
7 × 2 = 14	7 × ★ = 14 ★ = 2	14 ÷ 7 = 2
8 × 2 = 16	8 × ★ = 16 ★ = 2	16 ÷ 8 = 2
9 × 2 = 18	9 × ★ = 18 ★ = 2	18 ÷ 9 = 2
10 × 2 = 20	10 × ★ = 20 ★ = 2	20 ÷ 10 = 2

2. <u>Divide</u> and <u>write</u> the remainder in the parentheses.

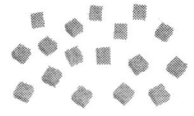

17 ÷ 2 = 8 (1)   17 ÷ 5 = 3 (2)
17 ÷ 7 = 2 (3)   17 ÷ 8 = 2 (1)   17 ÷ 3 = 5 (2)
17 ÷ 9 = 1 (8)   17 ÷ 6 = 2 (5)   17 ÷ 4 = 4 (1)

3. Dugtrio triggered 4 earthquakes. Each time it dug 8 miles underground. <u>How many miles</u> did it dig in all?

Answer: 4 × 8 = 32 (miles).

Page 60

1. <u>Divide</u> and <u>write</u> the remainder in the parentheses.

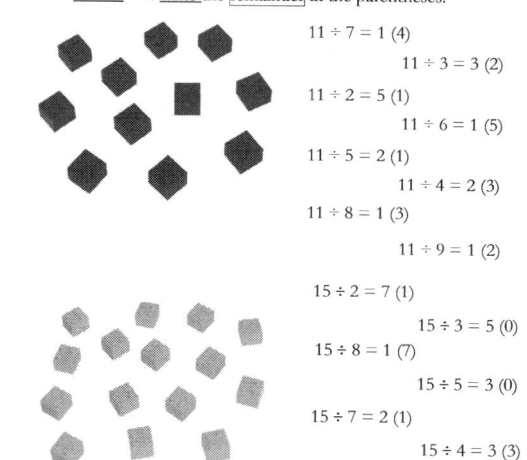

11 ÷ 7 = 1 (4)
11 ÷ 3 = 3 (2)
11 ÷ 2 = 5 (1)
11 ÷ 6 = 1 (5)
11 ÷ 5 = 2 (1)
11 ÷ 4 = 2 (3)
11 ÷ 8 = 1 (3)
11 ÷ 9 = 1 (2)

15 ÷ 2 = 7 (1)
15 ÷ 3 = 5 (0)
15 ÷ 8 = 1 (7)
15 ÷ 5 = 3 (0)
15 ÷ 7 = 2 (1)
15 ÷ 4 = 3 (3)
15 ÷ 9 = 1 (6)
15 ÷ 6 = 2 (3)

2. Meowth tried to capture 27 Pokemon in 27 attacks. It failed each 9th attack. <u>How many Pokemon</u> did it capture?

Answer: 27 − (27 ÷ 9) = 24 (Pokemon).

Page 61

1. <u>Multiply</u> and <u>divide</u>. <u>Connect</u> the multiplication and division diagrams of the same value.

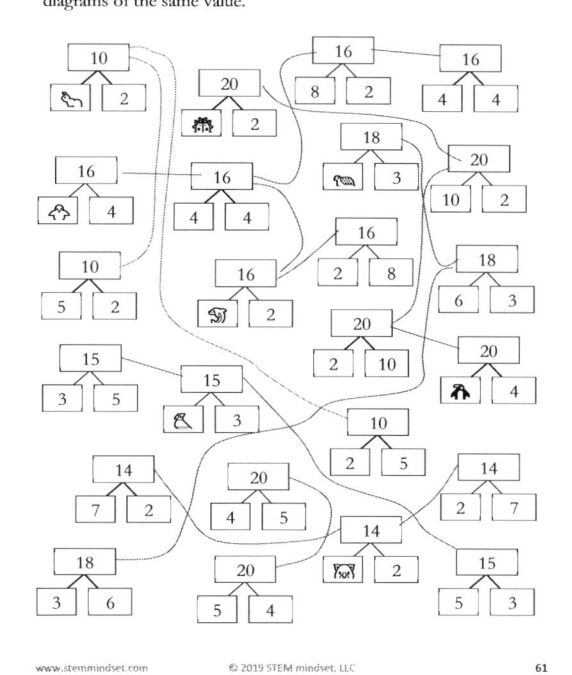

Page 62

1. <u>Divide</u> and <u>write</u> the remainder in the parentheses.

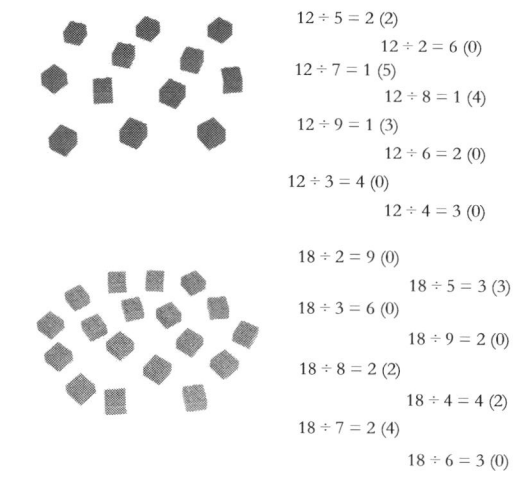

12 ÷ 5 = 2 (2)
12 ÷ 2 = 6 (0)
12 ÷ 7 = 1 (5)
12 ÷ 8 = 1 (4)
12 ÷ 9 = 1 (3)
12 ÷ 6 = 2 (0)
12 ÷ 3 = 4 (0)
12 ÷ 4 = 3 (0)

18 ÷ 2 = 9 (0)
18 ÷ 5 = 3 (3)
18 ÷ 3 = 6 (0)
18 ÷ 9 = 2 (0)
18 ÷ 8 = 2 (2)
18 ÷ 4 = 4 (2)
18 ÷ 7 = 2 (4)
18 ÷ 6 = 3 (0)

2. Seal uses its hard horn to smash through thick ice 3 times a day. <u>How many hits</u> did it use during the week?

Answer: 7 × 3 = 21 (hits).

Page 63

1. <u>Multiply</u> and <u>divide</u>. <u>Connect</u> the multiplication and division diagrams of the same value.

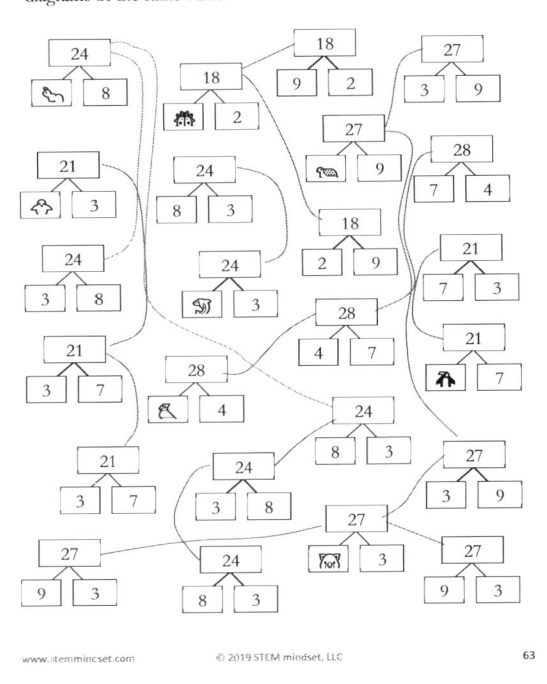

Page 64

1. <u>Divide</u> and <u>write down</u> the remainder in the parentheses.

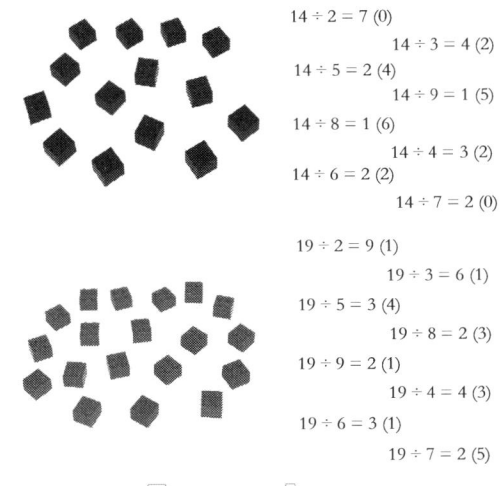

$14 \div 2 = 7 \ (0)$
$14 \div 3 = 4 \ (2)$
$14 \div 5 = 2 \ (4)$
$14 \div 9 = 1 \ (5)$
$14 \div 8 = 1 \ (6)$
$14 \div 4 = 3 \ (2)$
$14 \div 6 = 2 \ (2)$
$14 \div 7 = 2 \ (0)$

$19 \div 2 = 9 \ (1)$
$19 \div 3 = 6 \ (1)$
$19 \div 5 = 3 \ (4)$
$19 \div 8 = 2 \ (3)$
$19 \div 9 = 2 \ (1)$
$19 \div 4 = 4 \ (3)$
$19 \div 6 = 3 \ (1)$
$19 \div 7 = 2 \ (5)$

2. Horsea shot 24 blasts of ink in 6 battles. It failed each 2nd battle. <u>How many blasts of ink</u> reached the target?

$24 \div 6 = 4$ (blasts per battle)
$6 \div 2 = 3$ (battles)
Answer: $3 \times 4 = 12$ (blasts).

Page 65

1. <u>Get</u> Pikachu.

"Shame on you, Pikachu! Hiding from me! Where d'ya think yer going? Come back!"

2. <u>Find</u> and <u>circle</u> or <u>cross out</u> the words.

GRIMER
DIRT
MUK
SLIME
SLUDGE
TOXIC
POISONOUS
FOOTPRINTS
DAMAGE
FAILURE
HAZARD

Page 66

1. <u>Divide</u>.

$14 \div 7 = 2$	$20 \div 4 = 5$	$21 \div 3 = 7$
$12 \div 3 = 4$	$15 \div 5 = 3$	$16 \div 2 = 8$
$14 \div 2 = 7$	$20 \div 5 = 4$	$16 \div 8 = 2$
$12 \div 6 = 2$	$18 \div 3 = 6$	$20 \div 2 = 10$
$25 \div 5 = 5$	$18 \div 6 = 3$	$12 \div 2 = 6$
$18 \div 2 = 9$	$20 \div 10 = 2$	$16 \div 16 = 1$

2. <u>Complete</u> the addition number sentence for each multiplication number sentence and <u>find</u> the value.

$2 \times 6 = 6 + 6 = 12$ $5 \times 3 = 3+3+3+3+3=15$
$4 \times 2 = 2 + 2 + 2 + 2 = 8$ $6 \times 3 = 3+3+3+3+3+3=18$
$9 \times 3 = 3+3+3+3+3+3+3+3+3=27$ $7 \times 4 = 4+4+4+4+4+4+4=28$
$4 \times 6 = 6 + 6 + 6 + 6 = 24$ $8 \times 3 = 3+3+3+3+3+3+3+3=24$
$7 \times 3 = 3+3+3+3+3+3+3=21$ $5 \times 2 = 2+2+2+2+2=10$
$8 \times 4 = 4+4+4+4+4+4+4+4=32$ $6 \times 4 = 4+4+4+4+4+4=24$

3. Gyarados causes typhoons and sea storms while it gets angry. Last time it caused 3 typhoons and 4 times as many sea storms as typhoons. <u>How many times</u> did it get angry?

Answer: $(4 \times 3) + 3 = 15$ (times).

Pokemon Coloring Math Book Multiplication and Division Part 1 Grades 1-4 Ages 6+

1. <u>Divide</u> and <u>write</u> the remainder in the parentheses.

16 ÷ 2 = 8 (0)
16 ÷ 3 = 5 (1)
16 ÷ 5 = 3 (1)
16 ÷ 8 = 2 (0)
16 ÷ 4 = 4 (0)
16 ÷ 9 = 1 (7)
16 ÷ 7 = 2 (2)
16 ÷ 6 = 2 (4)

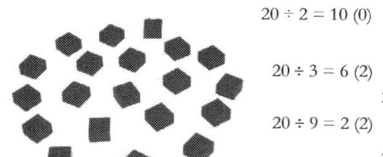

20 ÷ 2 = 10 (0)
20 ÷ 3 = 6 (2)
20 ÷ 5 = 4 (0)
20 ÷ 8 = 2 (4)
20 ÷ 9 = 2 (2)
20 ÷ 4 = 5 (0)
20 ÷ 6 = 3 (2)
20 ÷ 7 = 2 (6)

2. Draw a 4 × 8 forest where Bulbasaur lives.

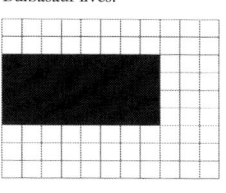

Pokemon Coloring Math Book Multiplication and Division Part 1 Grades 1-4 Ages 6+

1. <u>Evaluate</u> each expression.

18	16	14	20	12
÷ 2 9	÷ 4 4	÷ 2 7	÷ 10 2	÷ 2 6
- 7 2	- 2 2	+ 11 18	+ 14 16	- 5 1
× 4 8	× 10 20	÷ 9 2	÷ 4 4	× 8 8
÷ 1	÷ 5	× 6	÷ 2	÷ 4
8	4	12	2	2

20	20	15	12	9
÷ 4 5	÷ 5 4	÷ 5 3	÷ 4 3	÷ 3 3
- 2 3	- 2 2	+ 11 14	+ 13 16	- 2 1
× 4 12	× 9 18	÷ 7 2	÷ 4 4	× 15 15
÷ 6	÷ 6	× 6	× 5	÷ 3
2	3	12	20	5

2. <u>How many groups of 3</u> are in 15 Pikachu?   5

<u>How many groups of 9</u> are in 18 Pikachu?   2

<u>How many groups of 4</u> are in 16 Pikachu?   4

<u>How many groups of 2</u> are in 14 Pikachu?   7

Pokemon Coloring Math Book Multiplication and Division Part 1 Grades 1-4 Ages 6+

1. <u>Write</u> the missing numbers, <u>multiply</u> and <u>divide</u>.

4 × 3 = 12
12 ÷ 4 = 3
12 ÷ 3 = 4

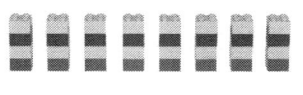

4 × 8 = 32
32 ÷ 4 = 8
32 ÷ 8 = 4

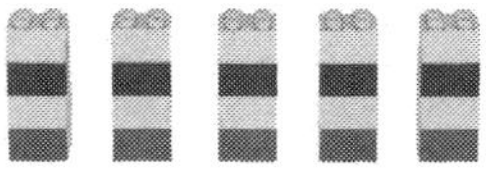

4 × 5 = 20
20 ÷ 4 = 5
20 ÷ 5 = 4

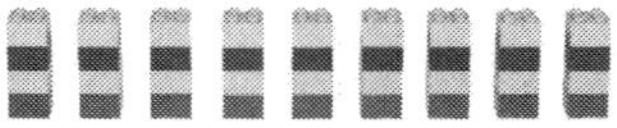

4 × 9 = 36
36 ÷ 4 = 9
36 ÷ 9 = 4

2. I have some numbers and signs: 2, 3, 2, ×, +.
<u>Write</u> the equation that equals one of the answer choices.

4 × 2 = 8
8 ÷ 4 = 2
8 ÷ 2 = 4

(A) 8   C   12
B   10   D   14

(2 × 3) + 2 = 8

Pokemon Coloring Math Book Multiplication and Division Part 1 Grades 1-4 Ages 6+

1. <u>Write</u> the missing numbers, <u>multiply</u> and <u>divide</u>.

4 × 4 = 16
16 ÷ 4 = 4
16 ÷ 4 = 4

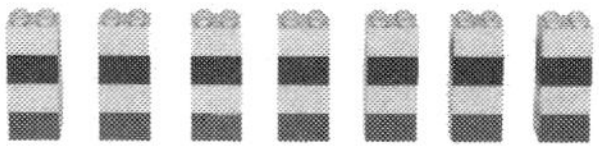

4 × 7 = 28
28 ÷ 4 = 7
28 ÷ 7 = 4

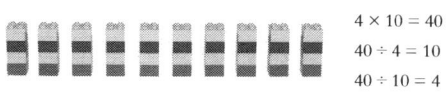

4 × 10 = 40
40 ÷ 4 = 10
40 ÷ 10 = 4

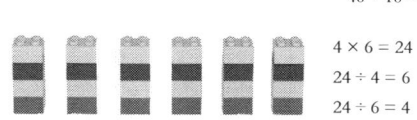

4 × 6 = 24
24 ÷ 4 = 6
24 ÷ 6 = 4

2. I have some numbers and signs: 7, 3, 5, ×, -.
<u>Write</u> the equation that equals one of the answer choices.

A   5   (C   6)
B   9   D   21

(7 - 5) × 3

Page 71

1. Multiply.

```
    3        5        8        1        4
  × 4      × 4      × 4      × 4      × 4
   1 2      2 0      3 2        4      1 6

    6        9        7        2       10
  × 4      × 4      × 4      × 4      × 4
   2 4      3 6      2 8        8      4 0
```

Speech bubble: "It's a marvelous little world with factors, products, quotients, mazes, and word search puzzles!"

Page 72

1. Make 4 rows. X shows the number of columns. Find X. Write the missing numbers, multiply and divide.

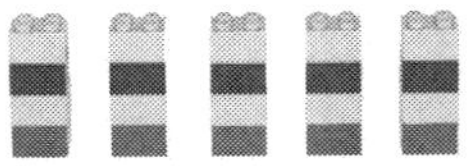

$4 \times X = 20$ $X = 5$
$20 \div 5 = 4$ $20 \div 4 = 5$
4 rows
5 bricks in each row

$4 \times X = 32$ $X = 8$
$32 \div 8 = 4$ $32 \div 4 = 8$
4 rows
8 bricks in each row

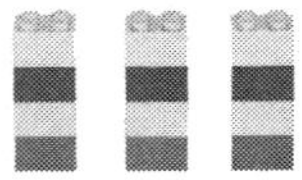

$4 \times X = 12$ $X = 3$
$12 \div 3 = 4$ $12 \div 4 = 3$
4 rows
3 bricks in each row

2. Circle the right answer.
A) $2 \times 5 \times 3$
B) $4 \times 2 \times 3$
(a) A is greater than B
b) A is equal to B
c) A is less than B

3. Circle the right answer.
A) $(4 \times 8) - 15$
B) $(5 \times 6) - 14$
(a) A is greater than B
b) A is equal to B
c) A is less than B

Page 73

1. Make 4 rows. X shows the number of columns. Find X. Write the missing numbers, multiply and divide.

$4 \times X = 40$ $X = 10$
$40 \div 10 = 4$ $40 \div 4 = 10$
4 rows   10 bricks in each row

$4 \times X = 16$ $X = 4$
$16 \div 4 = 4$ $16 \div 4 = 4$
4 rows
4 bricks in each row

$4 \times X = 28$ $X = 7$
$28 \div 7 = 4$ $28 \div 4 = 7$
4 rows
7 bricks in each row

2. Circle the right answer.
A) $3 \times 3 \times 3$
B) $1 \times 6 \times 5$
a) A is greater than B
b) A is equal to B
(c) A is less than B

Page 74

1. Make 4 rows. X shows the number of columns. Find X. Write the missing numbers, multiply and divide.

$4 \times X = 24$ $X = 6$
$24 \div 6 = 4$ $24 \div 4 = 6$
4 rows   6 bricks in each row

$4 \times X = 8$ $X = 2$
$8 \div 2 = 4$ $8 \div 4 = 2$
4 rows
2 bricks in each row

$4 \times X = 36$ $X = 9$
$36 \div 9 = 4$ $36 \div 4 = 9$
4 rows
9 bricks in each row

2. Circle the right answer.
A) $21 \div 7 \times 2$
B) $5 \times 4 \div 2$
a) A is greater than B
b) A is equal to B
(c) A is less than B

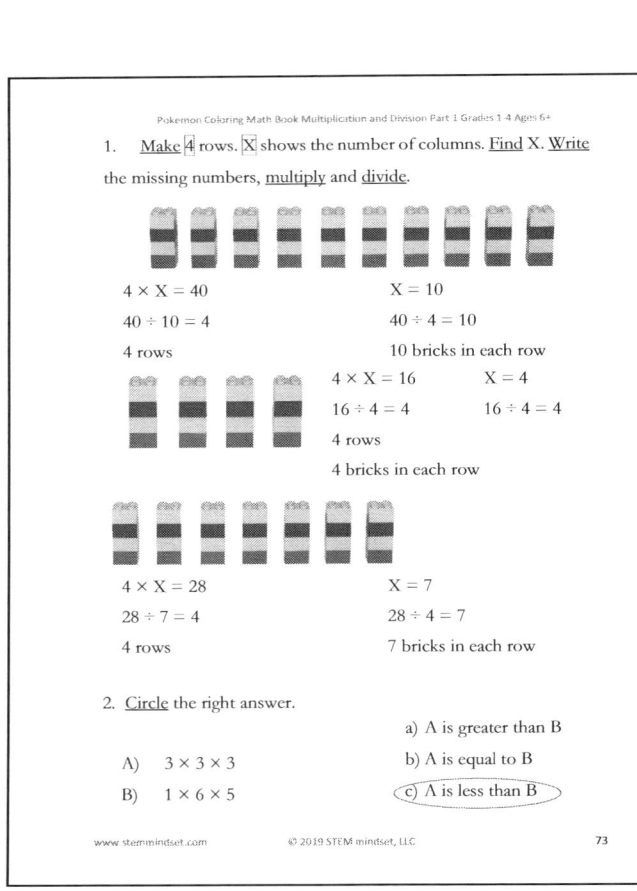

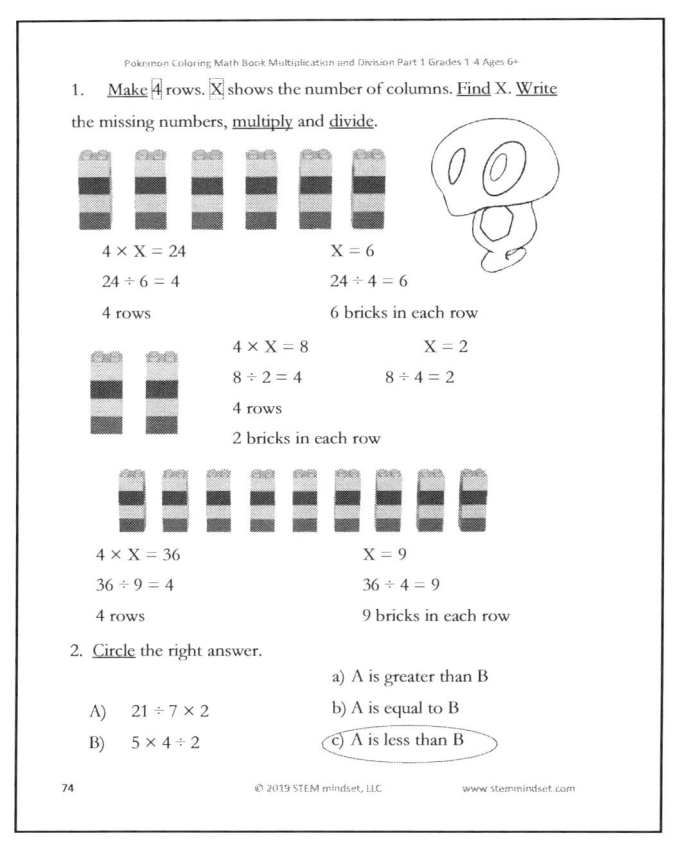

Page 75

1. <u>Write</u> the missing numbers.

1	3	6	10	9
× 4	× 4	× 4	× 4	× 4
4	12	24	40	36

4	7	2	5	8
× 4	× 4	× 4	× 4	× 4
16	28	8	20	32

I know my own strength! Pikachu, I'm coming!

Page 76

1. <u>Add</u> and <u>multiply</u>. The easiest way to count 4's is counting by groups of three 4's. Three 4's (4 + 4 + 4) equal 12.

4 + 4 + 4 = 12	3 × 4 = 12
4 + 4 + 4 + 4 = 16	4 × 4 = 16
4 + 4 + 4 + 4 + 4 = 20	5 × 4 = 20
+ 4 + 4 + 4 + 4 + 4 = 24	6 × 4 = 24
4 + 4 + 4 + 4 + 4 + 4 + 4 = 28	7 × 4 = 28
4 + 4 + 4 + 4 + 4 + 4 + 4 + 4 = 32	8 × 4 = 32
4 + 4 + 4 + 4 + 4 + 4 + 4 + 4 + 4 = 36	9 × 4 = 36
4 + 4 + 4 + 4 + 4 + 4 + 4 + 4 + 4 + 4 = 40	10 × 4 = 40

2. <u>Take</u> 36 bricks to answer the questions.

<u>Make</u> 4 rows.

<u>How many</u> columns can you make with 20 bricks?   5
<u>How many</u> columns can you make with the leftover bricks?   4
<u>How many</u> columns can you make out of all the bricks?   9.

Page 77

1. <u>Get</u> Pikachu.

Oh, goody, this maze is better than anything! Ho Ho ho ho – I'm glad I got Pikachu!

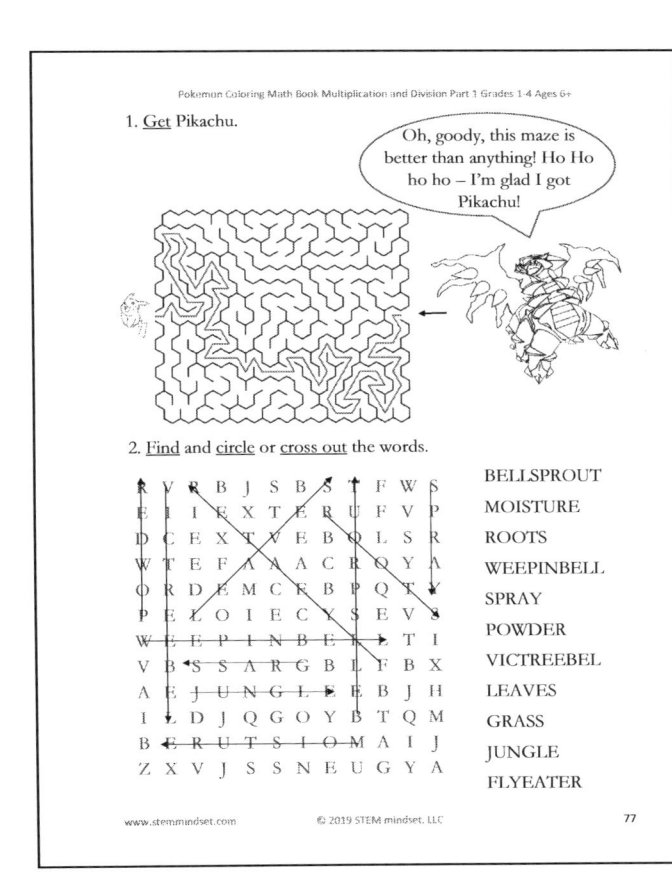

2. <u>Find</u> and <u>circle</u> or <u>cross out</u> the words.

BELLSPROUT
MOISTURE
ROOTS
WEEPINBELL
SPRAY
POWDER
VICTREEBEL
LEAVES
GRASS
JUNGLE
FLYEATER

Page 78

1. <u>Write</u> the missing numbers, <u>multiply</u> and <u>divide</u>.

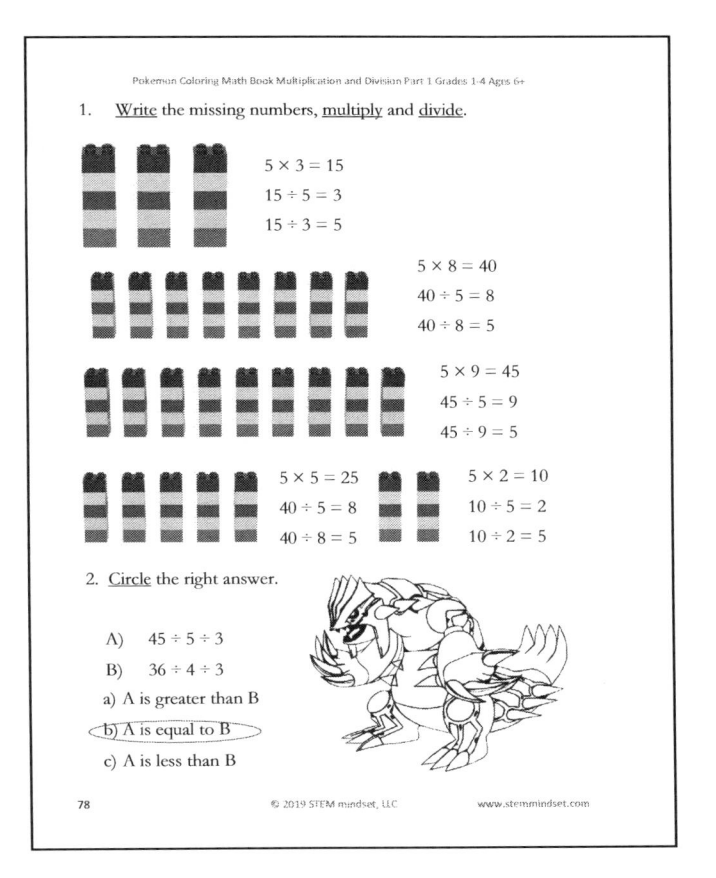

5 × 3 = 15
15 ÷ 5 = 3
15 ÷ 3 = 5

5 × 8 = 40
40 ÷ 5 = 8
40 ÷ 8 = 5

5 × 9 = 45
45 ÷ 5 = 9
45 ÷ 9 = 5

5 × 5 = 25   5 × 2 = 10
40 ÷ 5 = 8   10 ÷ 5 = 2
40 ÷ 8 = 5   10 ÷ 2 = 5

2. <u>Circle</u> the right answer.

A)   45 ÷ 5 ÷ 3
B)   36 ÷ 4 ÷ 3

a) A is greater than B
b) A is equal to B
c) A is less than B

Page 79

1. <u>Write</u> the missing numbers, <u>multiply</u> and <u>divide</u>.

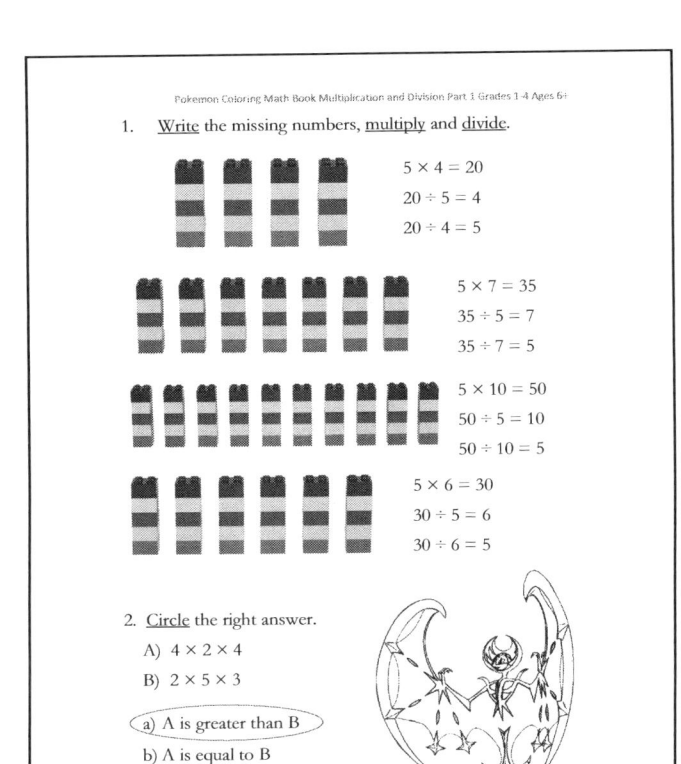

$5 \times 4 = 20$
$20 \div 5 = 4$
$20 \div 4 = 5$

$5 \times 7 = 35$
$35 \div 5 = 7$
$35 \div 7 = 5$

$5 \times 10 = 50$
$50 \div 5 = 10$
$50 \div 10 = 5$

$5 \times 6 = 30$
$30 \div 5 = 6$
$30 \div 6 = 5$

2. <u>Circle</u> the right answer.
A) $4 \times 2 \times 4$
B) $2 \times 5 \times 3$

(a) A is greater than B
b) A is equal to B
c) A is less than B

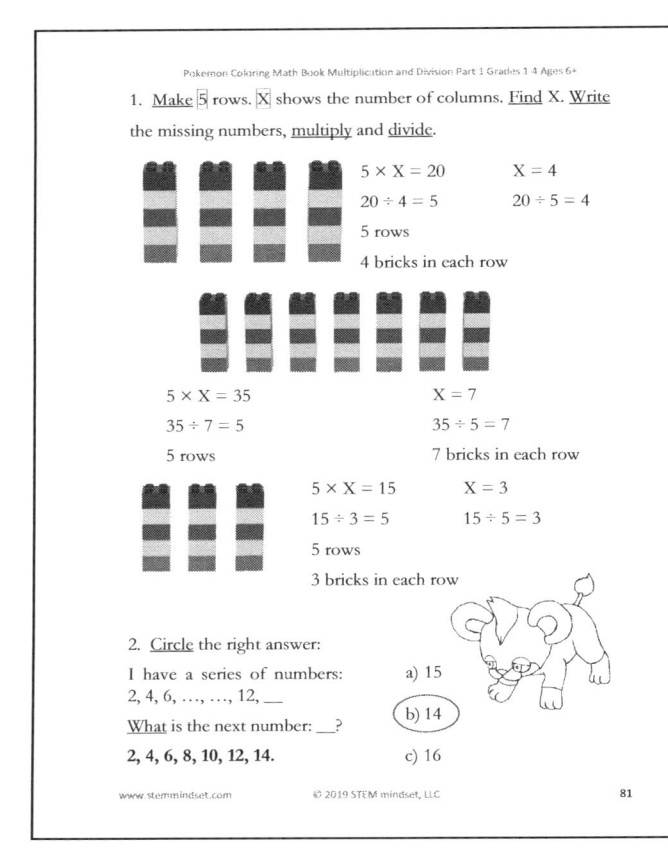

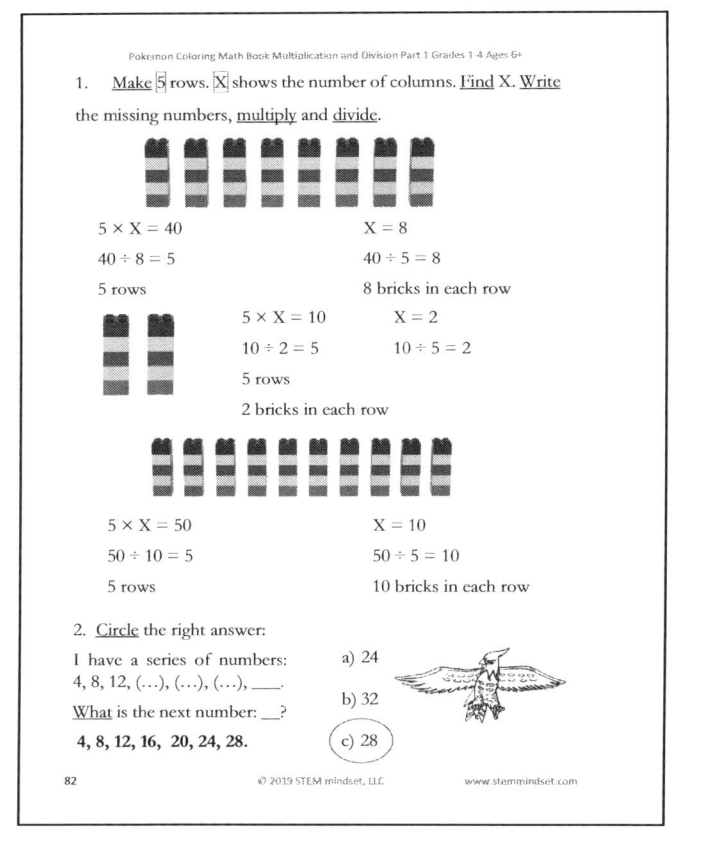

Page 80

1. <u>Multiply</u>.

3	5	8	1	4
$\times\ 5$	$\times\ 5$	$\times\ 5$	$\times\ 5$	$\times\ 5$
15	25	40	5	20

6	9	7	2	10
$\times\ 5$	$\times\ 5$	$\times\ 5$	$\times\ 5$	$\times\ 5$
30	45	35	10	50

"Poor little doomed Pikachu! Come, fella, let's leave now!"

Page 81

1. <u>Make</u> 5 rows. X shows the number of columns. <u>Find</u> X. <u>Write</u> the missing numbers, <u>multiply</u> and <u>divide</u>.

$5 \times X = 20$ $X = 4$
$20 \div 4 = 5$ $20 \div 5 = 4$
5 rows
4 bricks in each row

$5 \times X = 35$ $X = 7$
$35 \div 7 = 5$ $35 \div 5 = 7$
5 rows
7 bricks in each row

$5 \times X = 15$ $X = 3$
$15 \div 3 = 5$ $15 \div 5 = 3$
5 rows
3 bricks in each row

2. <u>Circle</u> the right answer:
I have a series of numbers:
2, 4, 6, …, …, 12, __
<u>What</u> is the next number: __?
2, 4, 6, 8, 10, 12, 14.

a) 15
(b) 14
c) 16

Page 82

1. <u>Make</u> 5 rows. X shows the number of columns. <u>Find</u> X. <u>Write</u> the missing numbers, <u>multiply</u> and <u>divide</u>.

$5 \times X = 40$ $X = 8$
$40 \div 8 = 5$ $40 \div 5 = 8$
5 rows
8 bricks in each row

$5 \times X = 10$ $X = 2$
$10 \div 2 = 5$ $10 \div 5 = 2$
5 rows
2 bricks in each row

$5 \times X = 50$ $X = 10$
$50 \div 10 = 5$ $50 \div 5 = 10$
5 rows
10 bricks in each row

2. <u>Circle</u> the right answer:
I have a series of numbers:
4, 8, 12, (…), (…), (…), __.
<u>What</u> is the next number: __?
4, 8, 12, 16, 20, 24, 28.

a) 24
b) 32
(c) 28

Page 83

1. Make 5 rows. X shows the number of columns. Find X. Write the missing numbers, multiply and divide.

$5 \times X = 45$ $X = 9$
$45 \div 9 = 5$ $45 \div 5 = 9$
5 rows   9 bricks in each row

$5 \times X = 25$ $X = 5$
$25 \div 5 = 5$ $25 \div 5 = 5$
5 rows
5 bricks in each row

$5 \times X = 30$ $X = 6$
$30 \div 6 = 5$ $30 \div 5 = 6$
5 rows   6
bricks in each row

2. I have some numbers and signs: 5, 8, 4, ×, ÷.
Make an equation that equals one the answer choices. Circle the right answer. $5 \times 8 \div 4 = 10$

A   3   C   8
(B) 10   D   6

Page 84

1. Write the missing numbers.

	1	4	7	3	5
×	5	5	5	5	5
	5	20	35	15	25

	8	10	2	9	6
×	5	5	5	5	5
	40	50	10	45	30

"Yoo Hoo, Pikachu! Do you mind telling us how you're solving all these mazes?"

Page 85

1. Find the value. The easiest way to count 5's is counting by groups of two 5's. Two 5's (5 + 5) equal 10.

$\underline{5 + 5} = 10$ $2 \times 5 = 10$
$\underline{5 + 5} + 5 = 15$ $3 \times 5 = 15$
$\underline{5 + 5} + \underline{5 + 5} = 20$ $4 \times 5 = 20$
$\underline{5 + 5} + \underline{5 + 5} + 5 = 25$ $5 \times 5 = 25$
$\underline{5 + 5} + \underline{5 + 5} + \underline{5 + 5} = 30$ $6 \times 5 = 30$
$\underline{5 + 5} + \underline{5 + 5} + \underline{5 + 5} + 5 = 35$ $7 \times 5 = 35$
$\underline{5 + 5} + \underline{5 + 5} + \underline{5 + 5} + \underline{5 + 5} = 40$ $8 \times 5 = 40$
$\underline{5 + 5} + \underline{5 + 5} + \underline{5 + 5} + \underline{5 + 5} + 5 = 45$ $9 \times 5 = 45$
$\underline{5 + 5} + \underline{5 + 5} + \underline{5 + 5} + \underline{5 + 5} + \underline{5 + 5} = 50$ $10 \times 5 = 50$

2. Take 45 bricks to answer the questions.

Make 5 rows.

How many columns can you make with 25 bricks?   5
How many columns can you make with the leftover bricks? 4
How many columns can you make out of all the bricks? 9

Page 86

"My-My-Who'd believe such a little monster capable of hiding so well!"

1. Find and circle or cross out the words.

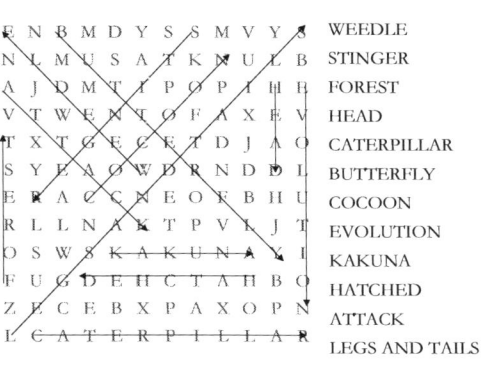

WEEDLE
STINGER
FOREST
HEAD
CATERPILLAR
BUTTERFLY
COCOON
EVOLUTION
KAKUNA
HATCHED
ATTACK
LEGS AND TAILS

Page 87

1. <u>Multiply</u>.

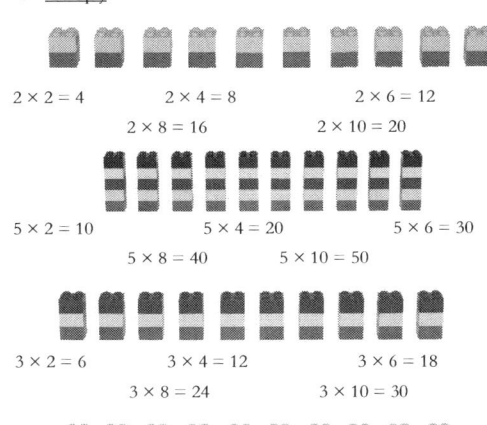

$2 \times 2 = 4$ $2 \times 4 = 8$ $2 \times 6 = 12$
$2 \times 8 = 16$ $2 \times 10 = 20$

$5 \times 2 = 10$ $5 \times 4 = 20$ $5 \times 6 = 30$
$5 \times 8 = 40$ $5 \times 10 = 50$

$3 \times 2 = 6$ $3 \times 4 = 12$ $3 \times 6 = 18$
$3 \times 8 = 24$ $3 \times 10 = 30$

$4 \times 2 = 8$ $4 \times 4 = 16$ $4 \times 6 = 24$
$4 \times 8 = 32$ $4 \times 10 = 40$

2. I have some numbers and signs: 6, 6, 9, ×, ÷.

<u>Make</u> an equation that equals one the answer choices. <u>Circle</u> the right answer. $6 \times 6 \div 9 = 4$

A 2   B 5   (C 4)   D 10

Page 88

1. <u>Multiply</u>.

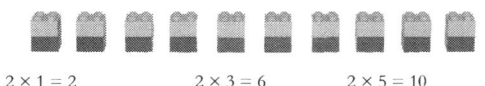

$2 \times 1 = 2$ $2 \times 3 = 6$ $2 \times 5 = 10$
$2 \times 7 = 14$ $2 \times 9 = 18$

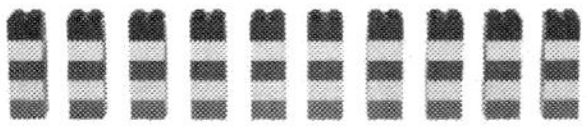

$5 \times 1 = 5$
$5 \times 3 = 15$ $5 \times 5 = 25$
$5 \times 7 = 35$ $5 \times 9 = 45$

$3 \times 1 = 3$ $3 \times 3 = 9$ $3 \times 5 = 15$
$3 \times 7 = 21$ $3 \times 9 = 27$

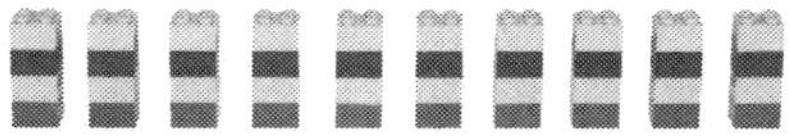

$4 \times 1 = 4$ $4 \times 3 = 12$ $4 \times 5 = 20$
$4 \times 7 = 28$ $4 \times 9 = 36$

Page 89

1. <u>Explain</u>: 12 is divided by 2. <u>Use</u> bricks, cubes, or sticks to answer the question.

It means <u>Answers will vary</u>: 12 is splitting into 2 equal parts or groups..

2. <u>Explain</u>: 24 is divided by 4. <u>Use</u> bricks, cubes, or sticks to answer the question.

It means <u>Answers will vary</u>: 24 is splitting into 4 equal parts or groups.

3. <u>Explain</u>: 36 is divided by 9. <u>Use</u> bricks, cubes, or sticks to answer the question.

It means <u>Answers will vary</u>: 36 is splitting into 9 equal parts or groups.

4. <u>Explain</u>: 35 is divided by 5. <u>Use</u> bricks, cubes, or sticks to answer the question.

It means <u>Answers will vary</u>: 35 is splitting into 5 equal parts or groups.

2. I have some numbers and signs: 2, 9, 3, ×, ÷.

<u>Make</u> an equation that equals one the answer choices. <u>Circle</u> the right answer. $2 \times 9 \div 3 = 6$

A 7   C 9
B 5   (D 6)

Page 90

1. <u>Evaluate</u> each expression. First, <u>start</u> with the parentheses (brackets), then, addition or subtraction from left to right! Or first, <u>start</u> with the parentheses (brackets), then, multiplication or subtraction. <u>Indicate</u> order of operations.

$(24 \div 3)^1 + (15 \div 3)^4 - (32 \div 4)^2 = 5$ $(18 \div 3)^1 \div (12 \div 4)^3 = 2$

$(20 \div 4) + (21 \div 3) - (12 \div 4) = 9$ $(27 \div 3) \div (9 \div 3) = 3$

$(12 \div 3) + (36 \div 4) - (18 \div 3) = 7$ $(28 \div 4) \times (8 \div 4) = 14$

$(27 \div 3) + (16 \div 4) - (9 \div 3) = 10$ $(20 \div 4) \times (16 \div 4) = 20$

$(28 \div 4) + (18 \div 3) - (8 \div 4) = 11$ $(36 \div 4) \times (6 \div 3) = 18$

$(21 \div 3) + (12 \div 4) - (24 \div 4) = 4$ $(32 \div 4) \div (12 \div 3) = 2$

$(35 \div 5) + (15 \div 5) - (32 \div 4) = 2$ $(45 \div 5) \div (12 \div 4) = 3$

$(25 \div 5) + (20 \div 4) - (10 \div 5) = 8$ $(30 \div 5) \div (8 \div 4) = 3$

$(30 \div 5) + (28 \div 4) - (45 \div 5) = 4$ $(40 \div 5) \times (16 \div 4) = 32$

$(35 \div 5) + (15 \div 5) - (18 \div 3) = 4$ $(45 \div 5) \div (27 \div 3) = 1$

$(25 \div 5) + (21 \div 3) - (10 \div 5) = 10$ $(30 \div 5) \div (9 \div 5) = 2$

$(30 \div 5) + (18 \div 3) - (45 \div 5) = 3$ $(40 \div 5) \times (12 \div 3) = 32$

Page 91

1. <u>Write</u> the missing numbers.

<u>Write</u> all even numbers from 1 up to 20.

2, 4, 6, 8, 10, 12, 14, 16, 18, 20

<u>Write</u> all odd numbers from 1 up to 20.

1, 3, 5, 7, 9, 11, 13, 15, 17, 19

<u>Write</u> all the numbers from 1 up to 20, divisible by 2.

2, 4, 6, 8, 10, 12, 14, 16, 18, 20

<u>Write</u> all the numbers from 1 up to 30, divisible by 3.

3,6,9,12,15,18,21,24,27,30

<u>Write</u> all the numbers from 1 up to 40, divisible by 4.

4,8,12,16,20,24,28,32,36,40

<u>Write</u> all the numbers from 1 up to 50, divisible by 5.

5, 10, 15, 20, 25, 30, 35, 40, 45, 50

Page 92

1. <u>Get</u> Pikachu.

Now, half the maze is yours, and half is mine — you take all the paths, I take Pikachu!

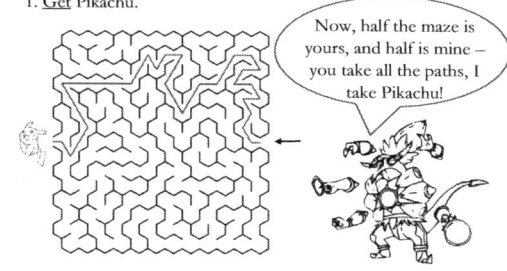

2. <u>Find</u> and <u>circle</u> or <u>cross out</u> the words.

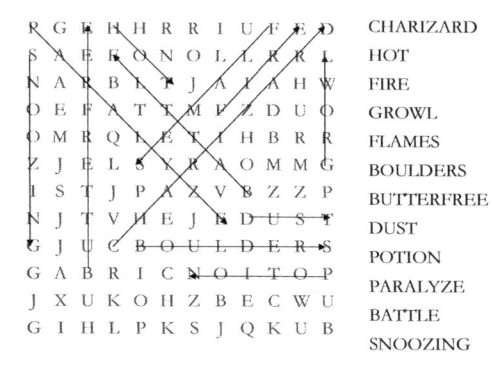

CHARIZARD
HOT
FIRE
GROWL
FLAMES
BOULDERS
BUTTERFREE
DUST
POTION
PARALYZE
BATTLE
SNOOZING

Page 93

1. <u>Multiply</u> and <u>divide</u>. Connect the multiplication and division diagrams of the same value.

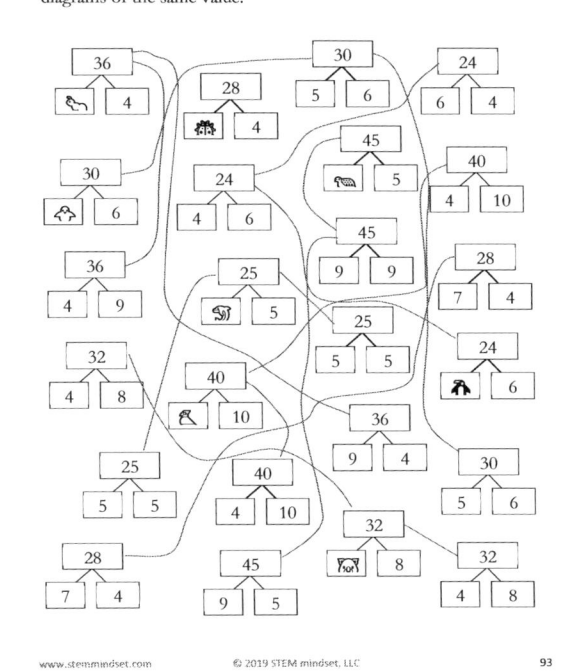

Page 94

1. <u>Complete</u> the addition number sentence for each multiplication number sentence and <u>evaluate</u> each expression.

$3 \times 4 = 4 + 4 + 4 = 12$ $8 \times 4 = 4+4+4+4+4+4+4+4=32$

$10 \times 2 = 2+2+2+2+2+2+2+2+2+2=20$ $4 \times 7 = 7+7+7+7=28$

$4 \times 4 = 4+4+4+4=16$ $3 \times 2 = 2+2+2=6$

$2 \times 9 = 9 + 9 = 18$ $5 \times 4 = 4+4+4+4+4=20$

2. <u>Evaluate</u> each expression. <u>Indicate</u> order of operations.

	1 2	1 2
$4 \times 5 = 20$	$(4 \times 3) + 23 = 35$	$(4 \times 5) - 8 = 12$
$4 \times 2 = 8$	$(4 \times 4) + 19 = 35$	$(4 \times 10) - 24 = 16$
$4 \times 10 = 40$	$(4 \times 7) + 6 = 34$	$(4 \times 9) - 18 = 18$
$4 \times 6 = 24$	$(4 \times 9) + 7 = 43$	$(4 \times 8) - 23 = 9$
$5 \times 5 = 25$	$(5 \times 3) + 15 = 30$	$(5 \times 6) - 24 = 6$
$5 \times 2 = 10$	$(5 \times 4) + 7 = 27$	$(5 \times 10) - 33 = 17$
$5 \times 10 = 50$	$(5 \times 6) + 24 = 54$	$(5 \times 9) - 17 = 28$
$5 \times 5 = 25$	$(5 \times 8) + 32 = 72$	$(5 \times 7) - 28 = 7$

1. <u>Multiply</u>.

5 × 4 = 20	5 × 8 = 40	5 × 2 = 10
5 × 5 = 25	5 × 10 = 50	5 × 1 = 5
2 × 4 = 8	2 × 8 = 16	2 × 2 = 4
2 × 5 = 10	2 × 10 = 20	2 × 1 = 2
4 × 4 = 16	4 × 8 = 32	4 × 2 = 8
4 × 5 = 20	4 × 10 = 40	4 × 1 = 4
3 × 4 = 12	3 × 8 = 24	3 × 2 = 6
3 × 5 = 15	3 × 10 = 30	3 × 1 = 3
5 × 3 = 15	5 × 7 = 35	5 × 9 = 45
2 × 3 = 6	2 × 7 = 14	2 × 9 = 18
4 × 3 = 12	4 × 7 = 28	4 × 9 = 36
3 × 3 = 9	3 × 7 = 21	3 × 9 = 27

---

1. <u>Divide</u>.

25 ÷ 5 = 5	50 ÷ 5 = 10	10 ÷ 5 = 2
20 ÷ 5 = 4	30 ÷ 5 = 6	40 ÷ 5 = 8
15 ÷ 5 = 3	35 ÷ 5 = 7	45 ÷ 5 = 9
10 ÷ 2 = 5	20 ÷ 2 = 10	4 ÷ 2 = 2
8 ÷ 2 = 4	12 ÷ 2 = 6	16 ÷ 2 = 8
6 ÷ 2 = 3	14 ÷ 2 = 7	18 ÷ 2 = 9
20 ÷ 4 = 5	40 ÷ 4 = 10	8 ÷ 4 = 2
16 ÷ 4 = 4	24 ÷ 4 = 6	32 ÷ 4 = 8
12 ÷ 4 = 3	28 ÷ 4 = 7	36 ÷ 4 = 9
15 ÷ 3 = 5	30 ÷ 3 = 10	6 ÷ 3 = 2
12 ÷ 3 = 4	18 ÷ 3 = 6	24 ÷ 3 = 8
9 ÷ 3 = 3	21 ÷ 3 = 7	27 ÷ 3 = 9

---

1. <u>Get</u> Pikachu.

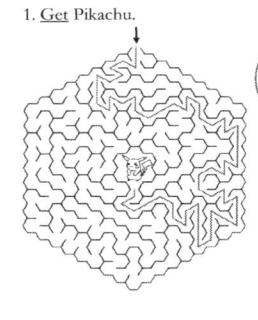

Yeh…I guess you're RIGHT about math, Pikachu! There's always plenty of fun in multiplying and dividing!

2. <u>Find</u> and <u>circle</u> or <u>cross out</u> the words.

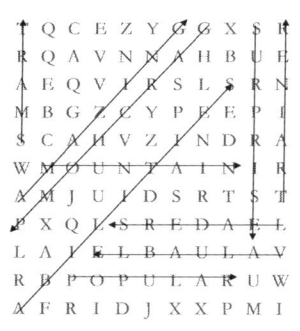

SMART
TRAINER
AMAZING
ABILITIES
POPULAR
MOUNTAIN
SURPRISE
GARCHOMP
VALUABLE
LEADERS

Made in the USA
Lexington, KY
15 November 2019